Cleanroom Design

Second Edition

Cleanroom Design

Second Edition

Edited by

W. WHYTE
University of Glasgow, UK

JOHN WILEY & SONS

Chichester • New York • Weinheim • Brisbane • Singapore • Toronto

Other Wiley Editorial Offices

John Wiley & Sons, Inc., 605 Third Avenue,
New York, NY 10158-0012, USA

Wiley-VCH Verlag GmbH
Pappelallee 3, D-69469 Weinheim, Germany

John Wiley & Sons Australia Ltd, 33 Park Road, Milton,
Queensland 4064, Australia

John Wiley & Sons (Asia) Pte Ltd, 2 Clementi Loop #02-01,
Jin Xing Distripark, Singapore 129809

John Wiley & Sons (Canada) Ltd, 22 Worcester Road,
Rexdale, Ontario, M9W 1L1, Canada

L:ibrary of Congress Cataloging-in-Publication Data

Cleanroom design / edited by W. Whyte. — 2nd ed.
p.   cm.
Includes bibliographical references and index.
ISBN 0-471-94204-9 (cloth. : alk. paper)
1. Cleanrooms.   I. Whyte, W.
TH7694.c54   1999
620.8—dc21
98-36213
CIP

British Library Cataloguing in Publication Data

A catalogue record for this book is available from the British Library

ISBN 0 471 94204 9

Typeset by Footnote Graphics, Warminster, Wiltshire
Printed and bound in Great Britain by Antony Rowe, Chippenham, Wiltshire
This book is printed on acid-free paper responsibly manufactured from sustainable forestry in which at least two trees are planted for each one used for paper production.

Contents

Contributors

Å. L. Möller
Cogito Consult AB, Angelholm, Sweden

J. G. King
King Consulting Incorporated, Eustace, Texas, USA

G. J. Farquharson
Tanshire Group of Companies, Elstead, UK

H. H. Schicht
Dr Hans Schicht Ltd, Zumikon, Switzerland

P. J. Tubito
Life Sciences International, Philadelphia, USA

T. J. Latham
Jacobs Engineering Ltd, Stockport, UK

S. D. Klocke
Flanders Filters, Washington, USA

E. C. Sirch
Bayer AG, Leverkusen, Germany

T. Hodgkiess
University of Glasgow, Glasgow, UK

R. Galbraith
Motorola Limited, Glasgow, UK

W. Whyte
University of Glasgow, Glasgow, UK

Preface

In recent years there has been a considerable expansion in the need for more and better cleanrooms. Many companies are using cleanrooms for the first time and other companies are developing new products that can only be produced in cleanrooms. The use of cleanrooms has mushroomed and a large range of products such as computers, CD players, lasers, navigational systems, medicines, medical devices, convenience foods, and various products of the new technologies, require a control of contamination that is only available in a cleanroom.

When I wrote the preface for the first edition of this book I said that the present cleanrooms are one million times cleaner than the original ones and that it was now possible to design a cleanroom to fulfil almost any contamination control requirement. I had therefore hoped that the first edition of the book could be used as the basis for cleanroom design into the foreseeable future. These words, that I wrote eight years ago, were a little optimistic. It has been necessary to bring out a second edition to update the information available. An additional chapter has been added to the book on the 'design of medical devices' and all of the other chapters have either been updated or completely rewritten. Some of the authors have changed but they still remain an international group from the United Kingdom, United States of America, Sweden, Germany and Switzerland, all with 20 to 35 years of experience. I trust that readers will agree that this book has benefited from the work that has gone into this second edition.

I would like to thank those who made the first and second edition of this book possible. It is now 35 years since I started to study cleanrooms. As a young man I was helped, and still am, by my colleague Bill Carson. In those early days, much of the fundamental work was carried out in hospital operating rooms and many of my ideas were shaped by research workers such as Dr Robert Blowers, Professor Edward Lowbury, the late Professor Sir John Charnley and, particularly Dr Owen Lidwell. I should also like to thank Sandra MacKay who did the initial typing and Barbara MacLeod who did the majority of the typing for the first edition and enthusiastically contributed to the editing of the first edition of the book. Isabelle Lawson produced many of the drawings used in this book.

W. Whyte
January, 1999

1 An Introduction to the Design of Clean and Containment Areas

W. WHYTE

INTRODUCTION

The cleanroom is a modern phenomenon. Although the roots of cleanroom design and management go back more than 100 year and are rooted in the control of infection in hospitals, the need for a clean environment for industrial manufacturing is a requirement of modern society. The use of cleanrooms is diverse and shown in Table 1.1 is a selection of products that are now being made in cleanrooms, or require contamination control facilities.

It may be seen that the requirement for cleanrooms can be broadly divided into two. The first area is that in which inanimate particles are a problem and where their presence, even in submicron size, may prevent a product functioning or reduce its useful life. The second group require the absence of microbe-carrying particles whose growth in the product (or in a hospital patient) could lead to human infection. It may also be seen that many of the examples given are recent innovations and this list will certainly be added to in the future, there being a considerable increase in the demand for these types of rooms.

THE HISTORY OF CLEANROOMS

It is clear that the first cleanrooms were in hospitals. The work of Pasteur, Koch, Lister and other pioneer microbiologists and surgeons over a hundred years ago established

TABLE 1.1. Some clean and containment room applications.

Electronics	Computers, TV tubes, Flat screens, Magnetic tape production
Semiconductors	Production of integrated circuits used in computer memory and control
Micromechanics	Gyroscopes, Miniature bearings, Compact disc players
Optics	Lenses, Photographic film, Laser equipment
Biotechnology	Antibiotic production, Genetic engineering
Pharmacy	Sterile pharmaceuticals
Medical devices	Heart valves, Cardiac by-pass systems
Food and drink	Disease-free food and drink
Hospital	Immunodeficiency therapy, Isolation of contagious patients, Operating rooms

Cleanroom Design. Edited by W. Whyte © 1999 John Wiley & Sons Ltd

that bacteria cause wound infections. It therefore followed that elimination of bacteria from the hospital and, in particular, the operating room should prevent infection. This was the scientific basis for the first cleanrooms. Lister substantially reduced infection in his operating room at the Royal Infirmary, Glasgow, by use of an antiseptic solution (carbolic acid) on the instruments, the wound and surgeon's hands and attempted to prevent airborne infection by spraying carbolic into the air. However his attempts at 'cleanliness' were by *antiseptics* and the progress to modern cleanrooms was through the path leading to the adoption of *aseptic* techniques, i.e. the sterilization of wound dressings, instruments and the use of surgical gloves, masks and gowns. These latter techniques are still the basis of many cleanroom techniques of today.

Although the cleanrooms of yesteryear had similarities to modern cleanrooms, a principal omission was positive ventilation by filtered air. The use of ventilation, albeit the natural type, to reduce bacterial infection had been advocated by people such as Florence Nightingale and mechanical ventilation was provided in the hospital designed for the Crimea by Brunel in 1855. However artificial ventilation was rare until about 60 years ago. Figure 1.1 shows that ventilation was used then to achieve comfort as well as hygiene and many designs advocated at that time, and later, for hospital ventilation did not clearly distinguish between the two. It is only towards the end of the Second World War that ventilation in hospitals was being clearly advocated for contamination control. The problems of airborne infection of people in crowded situations that occurred

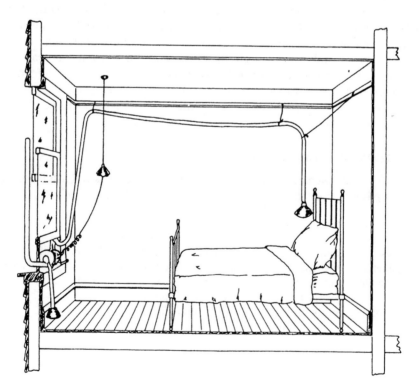

FIGURE 1.1. Ventilation of hospital room in the 1920s. A patient could inhale fresh air from the funnel. Foul air from the floor was extracted by another funnel.

in wartime, e.g. submarines, air raid shelters and army barracks, were studied. The ventilation of rooms with respect to infection, the invention of the bacterial slit sampler and the aerodynamics of particles were all studied during the Second World War. A great deal of early post-war work was carried out in operating and burns dressing rooms and by the early 1960s most of the principles which dictate the performance of conventional turbulently ventilated rooms were known, e.g. the patterns of air flow with respect to the type and placing of air diffusers and exhausts; the effect of a temperature difference between incoming and ambient air; the effect of air supply volume on the dilution of airborne contamination; air filter efficiency and air movement control between areas. Knowledge had advanced sufficiently well for a comprehensive guide to the design of an operating room ventilation system to be contained in a UK Medical Research Council Report published in 1962. Also established was the fact that people were the source of airborne bacteria, these being dispersed on rafts of skin, and that open weave cotton garments did little to prevent dispersion but tightly woven fabrics were required.

Directional airflow for the efficient removal of airborne contamination from hospital wards had been proposed by Sir John Simon, who wrote in 1864 that the ventilation must 'flow from inlet to outlet' and that this could only be achieved by a system of artificial supply whereby 'the currents are determinately regulated'. The Royal Victoria Hospital in Belfast, Northern Ireland, was built at the end of the previous century with a ventilation system intended to achieve this but the understanding of air movement was insufficient to produce the desired result.

Work by Bourdillon and Colebrook, which was published in 1946, described a dressing station in which there were 20 air changes per hour of filtered air which pressurized the room with respect to the outside areas. They also discussed the 'piston effect' in which air forms 'layers which are pushed down too slowly to cause a draught and which push the dirty air before them as they go'. They discussed the possibility of using 60 or greater air changes per hour and reported that laboratory tests gave a 'sudden disappearance of a bacterial cloud'. They did not go any further with this idea because of the cost of supplying such large quantities of air in a hospital.

This idea of downward displacement of air with a minimum of turbulence was further investigated by Blowers & Crew in 1960. They studied many aspects of ventilating an operating theatre, many of their ideas being adopted in the 1962 Medical Research Council report. At the suggestion of O. M. Lidwell they studied a room in which the air was supplied in a unidirectional manner by a diffuser across the whole ceiling. This system was the most effective of the many designs they investigated but once again it did not achieve its full potential because of the low air supply volume.

The main impetus for ultra-clean air for the operating room came from the work of Professor Sir John Charnley. Professor Charnley, in the early 1960s, radically improved the design and technique for insertion of an artificial hip joint. This is an extremely effective operation but the joint sepsis rate at the beginning of his studies was almost 9%. This was a disaster to the patients because, at that time, the methods used to treat these infections were not effective and the artificial joint had to be removed. Charnley thought that his infection probably arose from airborne bacteria. Assisted by Howorth Air Conditioning Ltd, he set about improving the air conditions in his theatre. He reported that to avoid turbulence and achieve a downward displacement at 60 ft/min (0.3 m/s) it would require, in an operating theatre with a ground plan of 20 ft × 20 ft

(6 m × 6 m), an air volume of 24 000 ft³ (11 m³/s). He considered this uneconomic and devised and installed in 1961 his sterile enclosure, known as a 'greenhouse', which was 7 ft × 7 ft (2 m × 2 m) in area. Shown in Figure 1.2 is his diagram of the air flow in the system. In his discussion of the work it is clear that he was dissatisfied with the downward displacement obtained and in June 1966, based on the experience of these prototypes, he installed an enclosure which gave substantially more air, much better downward displacement and hence much lower bacterial counts. He also invented a total-body exhaust gown to contain the bacteria dispersion from the surgeons. The infection rates through this series of improvements in ventilation fell from about 9% to 1.3% but because of the other advances in surgical techniques his claims, with regard to the effectiveness of ventilation, were disputed. However a study of the role of ultra-clean operating room systems was mounted in the 1980s by the Medical Research Council of the United Kingdom. Nineteen hospitals participated and the results were a vindication of Charnley's work. It was found that the use of unidirectional flow enclosures with occlusive clothing would reduce to a quarter the joint sepsis found in conventionally ventilated rooms.

In the engineering industries similar advances were being made. The development of the first cleanrooms for industrial manufacturing largely started during the Second World War in the United States and the United Kingdom mainly in an attempt to improve the quality and reliability of instrumentation used in guns, tanks and aircraft. It was realized that the cleanliness of the production environment had to be improved or such items as bomb sights and precision bearings would malfunction. At that time

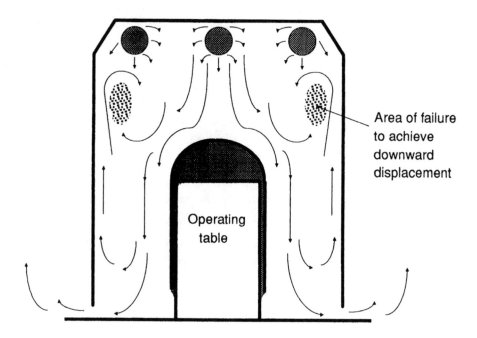

FIGURE 1.2. Enclosure showing flow or air from three diffuser bags (redrawn from article in *British Journal of Surgery*).

cleanrooms were built which copied operating room design and practices. However it was soon appreciated that 'bacteria-free' was not the same as 'particle-free'. A great deal of effort was therefore put into ensuring that materials and surfaces did not generate particles, but it was not fully appreciated that airborne dispersion of large quantities of particles by machines and people had to be removed by large quantities of pure air.

The peaceful and warfare uses of nuclear reaction as well as biological and chemical warfare research was the driving force for the production of high efficiency particulate air (HEPA) filters which were necessary to contain the dangerous microbial or radio-active contaminants. Their availability allowed cleanrooms to be supplied with clean air and low levels of airborne contamination to be achieved.

Rooms with large volumes of well filtered air supplied by ceiling diffusers were built between 1955 and the early 1960s, much of the impetus for the building of such rooms being the requirement for building inertial guidance systems.

The watershed in the history of cleanrooms was the realization in 1961 of the 'unidirectional' or 'laminar flow' concept of ventilation at the Sandia Laboratories, Albuquerque, New Mexico, USA. This was a team effort, but it is to Willis Whitfield that the main credit goes. Shown in Figure 1.3 is a drawing of their original uni-

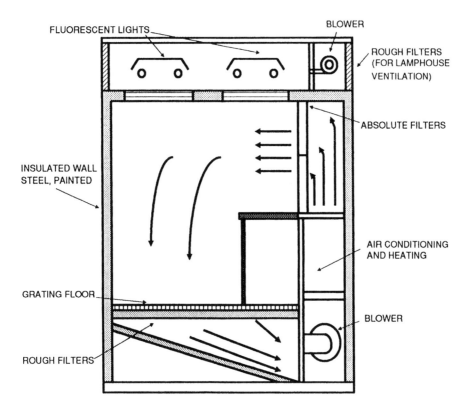

FIGURE 1.3. The original unidirectional flow cleanroom (redrawn from Sandia Corporation report).

directional air flow room. It was a small room 6 ft wide by 10 ft long by 7 ft high (1.8 m $\times$ 3 m $\times$ 2.1 m) and instead of the air being supplied by ceiling diffusers and moving about the room in a random way, it was supplied by a HEPA filter and moved in a unidirectional way from the filter across the room and out through the floor grille. It may be seen that anyone working at the bench in the room would not contaminate the work in front of them, as their contamination would be swept away. This concept of cleanroom ventilation was very quickly adopted into cleanrooms as higher quality cleanrooms were urgently required. In 1957 the USSR launched an orbiting satellite and the space race began. It was therefore important that a supply of miniaturized lightweight components be provided to lighten the payload. However, this miniaturization of components made them much more susceptible to particle contamination and it was quickly realized that very clean conditions were required and that unidirectional flow systems would give several magnitudes of cleanliness lower than previously achieved. This unidirectional method of ventilation was also quickly adopted by the pharmaceutical industry for the preparation of sterile products and in the hospital for the isolation of patients and the performing of surgical operations.

The Sandia team, aided by others from the US military, industry, and governmental agencies formed a group chaired by J. Gordon King (an author in this book) and produced in 1963 the first Federal Standard 209. This standard has had a major influence on the design of cleanrooms and is the basis of most world cleanroom standards. As can be read in Chapter 2 of this book, many standards have now been written to cater for the needs of the many areas of the expanded cleanroom industry.

The application of cleanrooms has increased and diversified. As well as minimizing the airborne contamination it may be necessary to contain dangerous or toxic contamination within the room. This is done by containment rooms.

Clean and containment rooms will be individually designed according to their application, but there are a number of basic similarities and design concepts that should be discussed before reading further chapters of this book. These concepts consider the special requirements of industries such as microelectronics, pharmaceuticals, medical devices and biotechnology.

CLEANROOMS

What is a Cleanroom?

It is clear that a cleanroom is a room that is clean. However, a cleanroom now has a special meaning and it is defined in Federal Standard 209E as:

> 'A room in which the concentration of airborne particles is controlled and which contains one or more clean zones.'

and in ISO 14644-1:

> 'A room in which the concentration of airborne particles is controlled, and which is constructed and used in a manner to minimize the introduction, generation, and retention of particles inside the room and in which other relevant parameters, e.g. temperature, humidity, and pressure, are controlled as necessary.'

Classification of Cleanrooms

Cleanrooms are classified by the cleanliness of their air. The method most easily understood and universally applied is the one suggested in versions of Federal Standard 209
up to edition 'D' in which the number of particles equal to and greater than 0.5 μm is
measured in one cubic foot of air and this count is used to classify the room.

A classification of cleanrooms according to the older Federal Standard 209D is given
in a simplified form in Table 1.2. This Federal Standard has now been superseded by

TABLE 1.2. A simplified Federal Standard 209D classification of cleanrooms.

Federal Standard 209 classification	1	10	100	1000	10 000	100 000
No. of particles/ft^3 ⩾ 0.5 μm	1	10	100	1000	10 000	100 000

a metric version (Federal Standard 209E) which was published in 1992. However,
because of its simplicity and universal use, it will be many years before the older
Federal Standard 209D classification is forgotten. It is also likely that Federal Standard
209D nomenclature will not be superseded by Federal Standard 209E but by the new
International Standard Organisation's (ISO) standard 14644-1.

The new ISO classification is based on the following equation:

$$C_n = 10^N \times \left[\frac{0.1}{D}\right]^{2.08}$$

where

C_n represents the maximum permitted concentration (in particles/m^3 of air) of
airborne particles that are equal to or larger than the considered particle size; C_n is
rounded to the nearest whole number,

N is the ISO classification number, which shall not exceed the value of 9. Intermediate ISO classification numbers may be specified, with 0.1 the smallest permitted increment of N,

D is the considered particle size in μm, and

0.1 is a constant with a dimension of μm.

The above equation was chosen so that the class limits set by the Federal Standard at
its standard reference point of 0.5 μm coincide closely with those found in the ISO
standard. This allows a harmonious transition from the previous cleanroom standards.

Reproduced in Figure 1.4 is the diagram given in the ISO standard which shows the
approximate class limits (maximum allowable airborne particle concentrations). These
are for illustration purposes only, as the precise limits are determined by the equation.
The ISO standard also gives a method by which cleanrooms can be specified in terms of
ultrafine particles (smaller than 0.1 μm) or macroparticles (larger than 5.0 μm).

Both the Federal Standard 209E and the new ISO standard are explained in much
more detail in the next chapter of this book. Throughout this book the classification of
cleanrooms will be given in terms of both the old Federal Standard 209D and the new
ISO standard nomenclature.

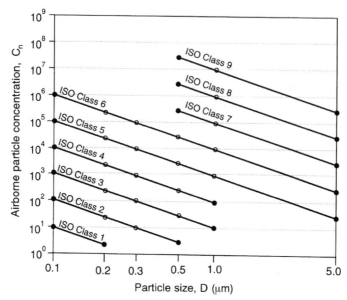

FIGURE 1.4. Representation of airborne particulate cleanliness classes according to the ISO standard.

It should be appreciated that the airborne contamination level of cleanroom is dependent on the particle-generating activities going on in the room. If a room is empty, very low particle concentrations can be achieved, these closely reflecting the quality of air supplied and hence the removal efficiency of the high efficiency filter. If the room has production equipment in it and operating, there will be a greater particle concentration but the greatest concentration will occur when the room is in full production. A classification of the room may therefore be carried out when the room is:

- as built: condition where the installation is complete with all services connected and functioning but with no production equipment, materials, or personnel present,
- at rest: condition where the installation is complete with equipment installed and operating in a manner agreed upon by the customer and supplier, but with no personnel present,
- operational: condition where the installation is functioning in the specified manner, with the specified number of personnel present and working in the manner agreed upon.

Class of Rooms Required by Different Industries

The required standard of cleanliness of a room is dependent on the task performed in it; the more susceptible the product is to contamination the better the standard. The following list (Table 1.3) gives an indication of the tasks carried out in different classifications of cleanrooms.

These suggested classifications are only an indication of what might be used and care must be taken not to over-design by providing cleaner than necessary rooms as this has a big influence on cost.

TABLE 1.3. Possible cleanroom requirement for various tasks carried out in cleanrooms.

Class 1	These rooms are only used by integrated circuit manufacturers manufacturing sub-micron geometries.
Class 10	These rooms are used by semiconductor manufacturers producing integrated circuits with line widths below 2 μm.
Class 100	Used when a bacteria-free or particulate-free environment is required in the manufacture of aseptically produced injectable medicines. Required for implant or transplant surgical operations. Isolation of immunosuppressed patients, e.g. after bone marrow transplant operations.
Class 1000	Manufacture of high quality optical equipment. Assembly and testing of precision gyroscopes. Assembly of miniaturized bearings.
Class 10 000	Assembly of prevision hydraulic or pneumatic equipment, servo-control valves, precision timing devices, high grade gearing.
Class 100 000	General optical work, assembly of electronic components, hydraulic and pneumatic assembly.

Types of Clean Areas

Clean areas can be divided into four main types. These are shown in a diagrammatic form in Figures 1.5–1.8 and are as follows:

1. *Conventional.* These cleanrooms are also known as turbulently-ventilated or non-unidirectional flow and are distinguished by their method of air supply. As can be seen in Figure 1.5, this is of the conventional type, the air being supplied by air supply diffusers or filters in the ceiling.
2. *Unidirectional flow.* This was previously known as laminar flow. As can be seen from Figure 1.6, clean air is supplied from a bank of high efficiency filters and passes in a unidirectional manner through the room.
3. *Mixed flow.* As shown in Figure 1.7, this type of cleanroom is conventionally ventilated but where the product is exposed to contamination, a unidirectional flow cabinet or workstation is used.
4. *Isolators or microenvironment.* These are used within a cleanroom to give the highest level of protection against contamination. In Figure 1.8 the isolator is shown to have a unidirectional supply of air but this may be a conventional turbulent-flow type. Similarly, gauntlets are shown, but half suites are also used.

Conventionally Ventilated Cleanrooms   Shown in Figure 1.9 is a diagram of a simple conventionally ventilated cleanroom. The general method of ventilation used in this type of cleanroom is similar to that found in offices, shops, etc. in that air is supplied by an air conditioning plant through diffusers in the ceiling. However, a cleanroom differs from an ordinary ventilated room in a number of ways:

1. *Increased air supply*: An office or shop will be supplied with sufficient air to achieve comfort conditions; this may be in the region of 2 to 10 air changes per hour. A typical conventionally ventilated cleanroom is likely to have between 20 and 60 air changes per hour. This additional air supply is mainly provided to dilute to an acceptable concentration the contamination produced in the room.

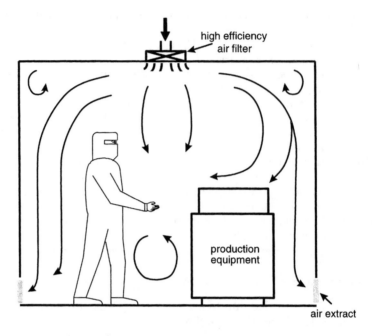

FIGURE 1.5. Conventional (non-unidirectional) airflow cleanrooms.

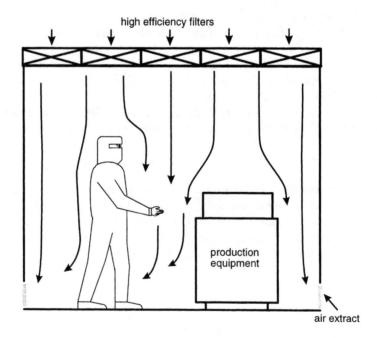

FIGURE 1.6. Vertical unidirectional flow cleanroom.

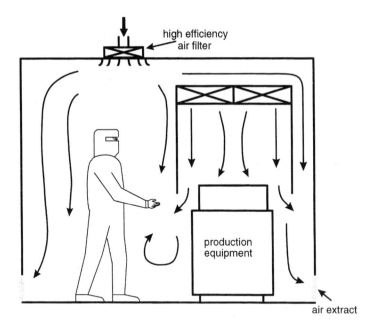

FIGURE 1.7. Mixed flow cleanroom with non-unidirectional flow in the room and unidirectional airflow protection for the critical processing area.

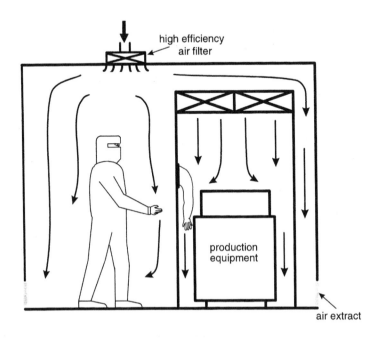

FIGURE 1.8. Isolator used to protect process area.

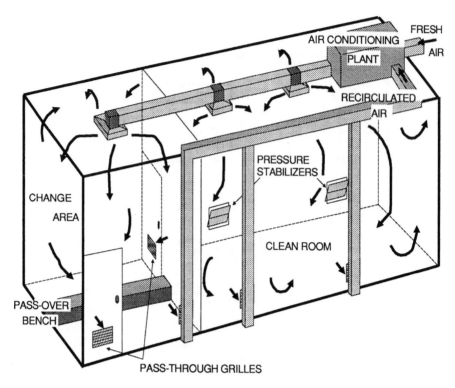

FIGURE 1.9. Conventionally ventilated cleanroom.

2. *High efficiency filters*: A cleanroom uses filters much more efficient than those used in offices etc. Cleanroom filters would normally be greater than 99.97% efficient in removing particles greater than 0.3 μm from the room air supply. These filters are known as High Efficiency Particle Air (HEPA) filters although Ultra Low Particle Air (ULPA) filters, which have a higher efficiency, are used in microelectronic fabrication areas.

3. *Terminal air filters*: The high efficiency filters used in cleanrooms are installed at the point of air discharge into the room. In air conditioning systems used in offices, etc. the filters will be placed directly after the ventilation plant but particles may be induced into the air supply ducts or come off duct surfaces and hence pass into the room.

4. *Room pressurization and pass-through grilles*: To ensure that air does not pass from dirtier adjacent areas into the cleanroom, the cleanroom is positively pressurized with respect to these dirtier areas. This is done by extracting less air from the room than is supplied to it, or by extracting the supplied air in adjacent areas. To achieve the correct pressure and allow a designed movement of air from the cleanest to the less clean rooms in a suite, pass-through grilles or dampers will usually be seen at a low level on walls or doors.

Another indication that the room is a cleanroom is the type of surface finish in a room. The room will be constructed of materials which do not generate particles and are easy to clean. Surfaces will be constructed so that they are accessible to cleaning and do not harbour dirt in cracks, e.g. coved flooring and recessed lighting.

The airborne cleanliness of a conventionally ventilated cleanroom is dependent on the amount and quality of air supplied to the room and the efficiency of mixing of the air. Generally speaking, a cleanroom will have sufficient air supply to achieve good mixing and the air quality of the room will therefore only depend on the air supply quantity and quality. It is important to understand that the cleanliness of a conventionally ventilated cleanroom is dependent on the volume of air supplied per unit of time and not the air change rate.

The cleanliness is also dependent on the generation of contamination within the room. i.e. from machinery and individuals working in the room. The more people in the cleanroom, the greater their activity and the poorer their cleanroom garments the more airborne contamination is generated. People moving about with poor cleanroom garments such as smocks or laboratory coats will generate, on average, about 2×10^6 particles $\geqslant 0.5$ μm/min, about 300 000 particles $\geqslant 5.0$ μm/min, and about 160 bacteria-carrying particles per minute. If people wear well designed clothing (coverall, knee-length boots, hood, etc.) made from tightly woven cloth the reduction of particles $\geqslant 0.5$ μm, $\geqslant 5.0$ μm and bacteria-carrying particles will be about 50%, 88% and 92%, respectively. Little information is available about the generation of particles from machinery used in cleanrooms but this may account for hundreds to millions of particles $\geqslant 0.5$ μm being dispersed per minute.

If the efficiency of the supply filters can be assumed to be close to 100% in removing the airborne contamination being considered, a rough approximation of the likely airborne cleanliness of a conventionally ventilated cleanroom (not a unidirectional flow one) can be achieved by use of the following equation:

$$\frac{\text{Airborne concentration}}{(\text{count/ft}^3 \text{ or m}^3)} = \frac{\text{Number of particles (or bacteria) generated/min}}{\text{Air volume supplied* (ft}^3 \text{ or m}^3\text{/min)}}$$

* Including that from unidirectional flow work stations.

Cleanrooms ventilated in this conventional turbulent manner may achieve conditions as low as ISO 6 (Class 1000) during manufacturing but are more likely to be ISO 7 (Class 10 000). To obtain cleaner rooms, greater dilution of the particles generated is necessary and this can be achieved by a unidirectional flow of air.

Unidirectional Airflow Cleanrooms   Unidirectional air flow is used when low airborne concentrations of particles of bacteria are required. This type of cleanroom was previously known as 'laminar flow', usually horizontal or vertical, at a uniform speed of between 0.3 and 0.45 m/s (60 to 90 ft/min) and throughout the entire air space.

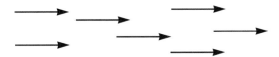

The air velocity suggested is sufficient to remove relatively large particles before they settle onto surfaces. Any contaminant generated into the air can therefore be immediately removed by this flow of air, whereas the conventional turbulently ventilated system relies on mixing and dilution to remove contamination. In a theoretical situation in an empty room with no obstructions to the airflow, contamination could be quickly removed to the exhaust by air velocities much lower than those mentioned above. However in a practical situation there are obstructions and people moving about. Obstructions will cause the unidirectional flow to be turned into turbulent flow and air vortexes to be established around the obstructions. Movement of people will also turn unidirectional into turbulent flow. Higher contamination concentrations will be established in these turbulent areas. It is therefore necessary that the velocity is in the region of 0.3 to 0.45 m/s (60 to 90 ft/min) so that the disrupted unidirectional flow can be quickly reinstated and the contamination around the obstructions be adequately diluted.

Unidirectional airflow is correctly defined in terms of air velocity, the cleanliness of a unidirectional room being directly proportional to the air velocity. Air changes per unit of time should not be used with a unidirectional flow room as they are related to the volume of the room, which generally has no effect on the performance of the system.

The air volumes supplied to unidirectional flow rooms are many times (10–100) greater than those supplied to a conventionally ventilated room. They are therefore very much more expensive in capital and running costs.

Unidirectional flow rooms are of two general types, namely horizontal or vertical flow. In the horizontal system the air flow is wall-to-wall and in the vertical system it flows from ceiling to ceiling.

Shown in Figure 1.10 is a typical vertical flow type of cleanroom. It may be seen that the air is supplied from a complete bank of high efficiency filters in the roof and this flows vertically through the room and out through open grilled flooring. Air in this figure is shown to flow through the complete area of a floor but it is common to find rooms in which the air returns through grilles which are distributed about the floor. If the floor area is not too great, grilles can alternatively be placed at a lower level in the

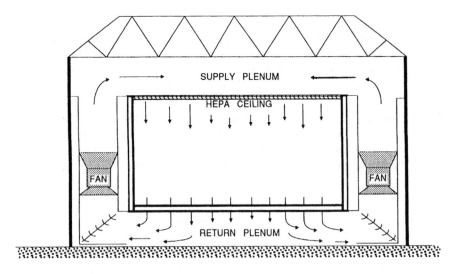

FIGURE 1.10. Vertical unidirectional flow cleanroom.

walls. The exhaust air is recirculated, mixed with some fresh make-up air, and supplied to the room through the high efficiency filters in the room ceiling.

Most unidirectional cleanrooms are built in a vertical manner as particles generated within the room will be quickly swept down and out of the room. Less popular is the horizontal flow type of cleanroom, a typical example being shown in Figure 1.11. This type of cleanroom is not so popular because any contamination generated close to the filters will be swept down the room and could contaminate work processes downwind. However as the area of a wall in a room is usually much smaller than the ceiling the capital and running costs are less. If the cleanroom can be arranged so that the most critical operations are close to the supply filters and the dirtier ones at the exhaust end, then this type of room can be successful. One such application is used in the nursing of patients susceptible to microbial infection, where the patient's bed is placed next to the filters and doctors and nurses can perform many of their tasks downwind (Figure 1.12).

Mixed Flow Rooms This type of room is a conventional flow room in which the critical manufacturing operations are carried out within a higher quality of air provided by a unidirectional flow system, e.g. a bench. This mixed type of system is very popular as the best conditions are provided only where they are needed and considerable cost savings are available for use in this room. Shown in Figure 1.13 is a horizontal flow cabinet, this being one of the simplest and most effective methods of controlling contamination. In this bench the operator's contamination is kept downwind of the critical process. Also available are a variety of styles of vertical flow systems which may vary in size to encompass a person's manipulations or large pieces of machinery.

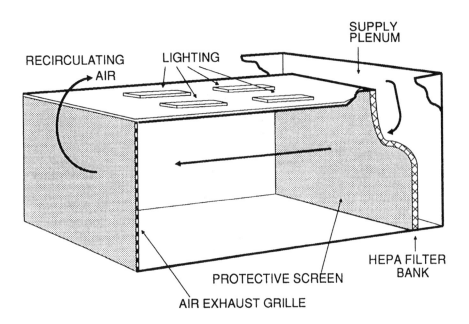

FIGURE 1.11. Horizontal unidirectional flow cleanroom.

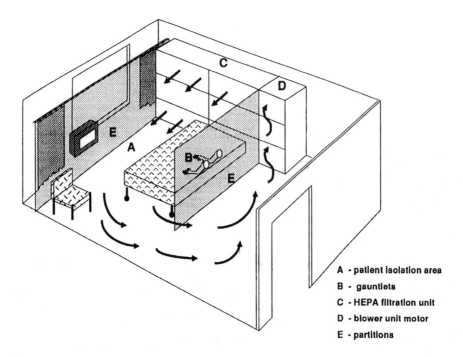

A - patient isolation area
B - gauntlets
C - HEPA filtration unit
D - blower unit motor
E - partitions

FIGURE 1.12. Horizontal unidirectional flow room for a hospital.

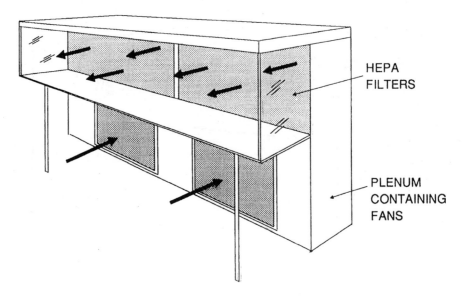

HEPA
FILTERS

PLENUM
CONTAINING
FANS

FIGURE 1.13. Horizontal flow cabinet.

Isolator or Minienvironment   Hazardous work with toxic chemicals or dangerous bacteria has been carried out for many years in glove boxes. Work on germ-free animals has also been carried out for decades in plastic isolators which prevented the entrance of microorganisms. These contaminant-retaining and contaminant-excluding systems do not principally depend on airflow for isolation but walls of metal and plastic. This principle of isolation clearly has excellent barrier properties and it has now been developed for use in modern cleanroom technology. In the pharmaceutical manufacturing area this technology is generally known as isolator or barrier technology, whereas in the semiconductor industry it is generally known as minienvironments.

Shown in Figure 1.14 is a system of interlocked plastic film isolators of the type used in pharmaceutical manufacturing. It may be seen that the plastic sheet acts as a barrier to outside contamination, and personnel either enter into half suits or use gauntlets to work at the clean processes within the isolators. The air within the isolator is sterile and particle-free having been filtered by high efficiency filters and this air is also used to pressurize the system and prevent the ingress of outside contamination. In this system the containers enter the isolator system through a sterilizing tunnel. They are then filled with liquid, inspected, freeze dried, and capped in the various isolators.

Another system, which is used in semiconductor manufacturing, is the SMIF (Standard Mechanical Interface Format) system. In this system silicon wafers are transported between machines in special containers which prevent the wafers being contaminated by the air outside. These containers, which contain the wafers, are slotted into the machine interface, the wafers processed and then loaded onto another container which can be taken to another machine and loaded into its interface. This system is further described and discussed in Chapter 3.

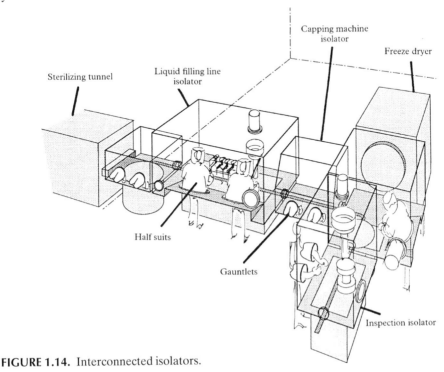

FIGURE 1.14. Interconnected isolators.

CONTAINMENT OF CONTAMINATION

Cleanrooms are used to prevent contamination of articles produced in the room. However it is quite common to find that some manufacturing processes produce toxic chemicals or dangerous bacteria and these must be contained. This can occur, for example, in the pharmaceutical industry where highly active pharmaceuticals, such as hormones, must not reach the operator. Other examples are to be found in the biotechnology industry where rooms are required to contain the genetically engineered microorganisms. Microbiological laboratories dealing with disease-producing microorganisms require to ensure that the personnel working in them, or the people passing near them, are not infected.

The technology associated with the design of these containment rooms is similar to that used in cleanrooms and it is normal that containment rooms should also be cleanrooms. It is also common to find cleanrooms with containment cabinets within them.

Containment Rooms and Cabinets

Shown in Figure 1.15 is an example of a containment room that might be used for working with microorganisms dangerous to the health of the personnel working there, or to anyone passing close to the room. It may be seen that clean air is supplied to the room but more air must be extracted from the room so that the room will be under negative pressure and air will always flow into the room. The air that is extracted must be filtered through a high efficiency HEPA filter before being discharged to the outside.

Within this room there will be a safety cabinet in which the microorganisms are manipulated. In a room where there is not a very high risk a Class I or Class II cabinet is used. In a high-risk area a Class III cabinet would be used. Shown in Figure 1.16(a), (b) and (c) is a diagrammatic representation of these three types of cabinets showing their air flow and isolation principles. If the manipulation in the cabinet requires clean conditions then a Class II cabinet will be used, as this type is designed to give a flow of filtered air over the product while still ensuring that the flow of air is into the cabinet. To ensure satisfactory working of a class II cabinet, attention must be routinely given

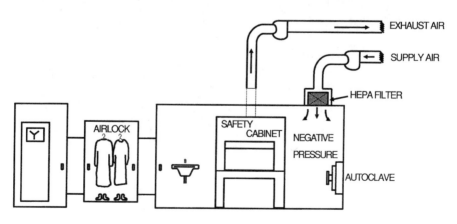

FIGURE 1.15. Containment room.

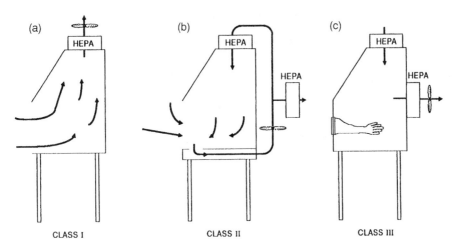

FIGURE 1.16. Containment cabinets.

to the air flow balance and if especially clean conditions are not necessary a Class I cabinet may be chosen for its more stable airflow balance.

Other features which may be seen in such rooms are the use of an airlock to allow people to pass in and out of the room. A pass-through autoclave may be available to allow for the sterilization of contaminated material.

Other containment rooms may be of a higher or lower standard, depending on the toxic, chemical, or microbiological hazard of the room. Less hazardous rooms would not use an airlock or pass-through autoclave and rely on the exhaust of a Class I cabinet, or fume cupboard, to create a negative pressure in the room. Rooms in which the hazard was high would contain the hazard within a Class III type of cabinet and provide a shower area between the airlock and the room. In particularly hazardous situations, filtered air suits would be worn by personnel.

SUPPLY OF LIQUID AND GASES TO CLEANROOMS

To ensure that the product produced in the room is free of particles and microbial contamination, it is necessary to ensure that not only the air is free of contamination but that other gasses and liquids supplied to the room are also free of contamination. In pharmaceutical cleanrooms there is the requirement for large quantities of water used to make up pharmaceuticals and in semiconductor fabrication areas pure water is used to wash silicon wafers during the manufacture. The manufacture of semiconductors also requires the supply of various gases and these must be provided with extremely low levels of contamination. These topics are discussed in other chapters of this book.

ACKNOWLEDGEMENTS

Figure 1.2 is redrawn from the *British Journal of Surgery* with the permission of the publishers, Butterworth and Co. Figure 1.14 is reproduced by permission of La Calhene Ltd.

BIBLIOGRAPHY ON THE HISTORY OF CLEANROOMS

Blowers, R. and Crew, B. (1960). 'Ventilation of operating theatres', *Journal of Hygiene, Cambridge*, **58**, 427–448.

Bourdillon, R. B. and Colebrook, L. (1946). 'Air hygiene in dressing-rooms for burns or major wounds', *Lancet*, (i), 601–605.

Charnley, J. (1964). 'A sterile-air operating theatre enclosure', *British Journal of Surgery*, **51**, 195–202.

King, J. G. (1986). 'The history of clean rooms', *ASHRAE Transactions*, **92** (1B), 299.

Lidwell, O. M. (1987). 'Joseph Lister and infection from the air', *Epidemiology and Infection*, **99**, 569–578.

Medical Research Council Report. (1962). 'Design and ventilation of operating-room suites for control of infection and for comfort', *Lancet*, (ii), 945–951.

Peck, R. D. (1981). 'History and present status of contamination control standards', *Proceedings of the 27th Annual Technical Meeting*. Institute of Environmental Sciences, USA, pp. 57–63.

Whitfield, W. J. (1962). 'A new approach to cleanroom design', *Sandia Corporation Report* SC-4673 (RR), Office of Technical Services, Department of Commerce, Washington 25, DC, USA.

Whitfield, W. J. (1981). 'A brief history of laminar flow clean room systems', *Proceedings of the 27th Annual Technical Meeting*. Institute of Environmental Sciences, USA, pp. 15–17.

2 International Standards for the Design of Cleanrooms

Å. L. MÖLLER

INTRODUCTION

This chapter deals with the standards, norms, orders, practices and recommendations that are available internationally and should be considered when designing a cleanroom. These standards are presented in this chapter in three groups, namely

- general engineering standards
- pharmaceutical or biocontamination standards, and
- containment standards

The current standards pertaining to these above groups are described, together with a number of other useful older standards. However, it should be noted that cleanroom standards are being developed for international use by the International Standards Organization (ISO), and for use within the European Union by the Comité Européen de Normalization (CEN). Information about this on-going work is also given within this chapter.

As well as information on standards, there is a list of references of monographs, standards and guidelines given in this chapter, as well as a list of abbreviations and the address of the sources of the documents discussed in this chapter.

The aim of this chapter is therefore to describe well-known, established standards and to inform the reader about forthcoming international standards, which should be considered when designing cleanrooms.

THE CLEANROOM STANDARDS

The Naming of Standards, Practices and Technical Orders

In this chapter the word *standard* is used as a general term for words of a similar meaning. In some cases an official national, or international, standard exists, while in other cases the authorities have only produced a recommendation, a specification, or a guide. There are also internal standards or orders. Such standards or orders can, however, be used internationally, as in the case of US Military Technical Orders.

The most well-known and widely used cleanroom standard in the world, which was

Cleanroom Design. Edited by W. Whyte © 1999 John Wiley & Sons Ltd

originally issued in 1963, is the United States Federal Standard 209. When this was first published it was not a complete standard in the accepted definition of the word, as a major part of it was not mandatory. However, the latest version of it, 209E, which was issued in September 1992, is a complete standard, but has a much more limited scope than the previous versions.

Standards written for the design of cleanrooms are part of a more comprehensive group of contamination control standards. There is a comprehensive survey of the many contamination control standards that are available and this was published by the Institute of Environmental Sciences and Technology (IEST) in the United States and known as the 'Compendium of Standards, Practices, Methods and Similar Documents, Relating to Contamination Control; IEST-CC-009'. The 1993 version covers 276 standards, 143 from outside the United States. This document, IEST RP 009.2, has a helpful category listing and gives full addresses of where to buy the listed documents. This revised edition is much more complete than the earlier one, but we must consider that a list of all cleanroom standards will never be completed, as new or revised cleanroom standards are always being developed.

During the last 5–10 years there were annually about 30 new or revised cleanroom standards presented in the major industrial nations, covering different applications and written in a dozen different languages. This rate of expansion will now be reduced due to the CEN and ISO work for common, international standards.

Who Produces Cleanroom Standards?

Contamination control standards, up to the 1990s, were mainly written for use in the countries where they were written and much work was done by the national standards organizations in the United States, European countries and Japan (see the list at the end of this chapter).

National contamination control societies, of which there were 16 in 1998, also carried out the important work of writing standards. Their confederation, the International Confederation of Contamination Control Societies (ICCCS), has as one goal the promotion of international cleanroom standards. ICCCS is also a liaison partner in the ISO work .

However, some international organizations like ISO, FIP and WHO (see end of chapter for an explanation of the abbreviations), and European organizations like Eurovent and PIC, which covered about a dozen countries, also wrote standards on cleanrooms at that time. Recently, much work has been done on European standards by CEN in the European Union and by ISO for world-wide standards. Within a few years, this CEN and ISO work will completely change the structure of the major contamination control standards.

International Development of Cleanroom Standards

The Start   Hospitals were aware from an early date of the problems of contamination. Microorganisms had been known since the nineteenth century to cause infections. Between the 1940s and 1960s, studies were performed in hospitals to demonstrate the relationship between airborne microorganisms and infections, and cleanroom standards for hospitals were written in the early 1960s.

The development of industrial cleanroom standards goes back to the 1950s in the United States. The HEPA filter was developed for military use and military cleanroom standards were introduced. One such example is a standard written for the testing of HEPA filters by US Army Armament R&D dated 1950.

Room air in those earlier days was tested by sucking air through a membrane filter and counting the particles on the filter surface with a microscope. For the development of better cleanroom standards, better and faster counting instruments had to be developed, and in the late 1950s light-scattering instruments were made available. These were called Optical Particle Counters (OPCs), or today Discrete Particle Counters (DPCs), thus starting a new era of classifying clean air and cleanrooms.

During the 1960s the first steps were taken towards the development of cleanroom standards. The start of the space era played an important role; the launching of the 'Sputnik' in 1957 starting a space race between the USSR and the United States. The very important 'laminar' or unidirectional flow concept, which was developed at the Sandia Corporation, USA, was reported in January 1962. In December 1963 the first edition of the US Federal Standard 209 was published.

Before the production of the Federal Standard 209, several military (mainly Air Force) Standards and Technical Orders were published in the United States and other countries. In the early 1960s the Swedish Air Force introduced Technical Orders to avoid contamination problems. At the end of the 1960s the American Association for Contamination Control's standardization activity started in the United States with standards for HEPA filters, etc. Contamination control associations started in several countries, one of their aims being to develop cleanroom standards.

During the 1970s many cleanroom standards were written in the United States, as well as in Japan, the United Kingdom, France and Germany. European organizations like Eurovent and PIC were also engaged in cleanroom standards. National standards organizations were also active. Standards for occupational safety, safety classes (P1–P4), and biohazard equipment (safety benches Classes I–III) were also published during this period.

Today   Between 1960 and 1990, the number of cleanroom standards grew markedly in the United States, Europe and Japan. The earlier mentioned IEST survey of all cleanroom standards (RP-009), when published in 1993, reported 276 official standards, which was double the number of the 1984 edition. The number of cleanroom standards in the United States given in RP 009 is 167, that country therefore dominating the survey.

To avoid the confusion caused by the many similar standards produced in the individual countries and to overcome trade barriers in the growing international trade, it is necessary to have common international standards. On request from the British Standards Institution, the Comité Européen de Normalization (CEN), in 1990, started a Technical Committee, TC 243, to write European cleanroom standards. By 1994, seven preliminary European cleanroom standards had been produced.

However, in 1992, the Institute of Environmental Science and Technology in the United States finalized its revision to FS 209E and presented a proposal to the International Standards Organization (ISO) to create international cleanroom standards. After consulting its members, ISO set up a technical committee (TC 209) in November 1993. Work was suspended on the CEN standards and the work was carried over from the CEN committees and utilized in the ISO standards.

When the ISO standards are produced, CEN will have the option of adopting them as European Standards; this is very likely. When CEN standards are published, any similar official standards within individual countries of the EU will be superseded. ISO have groups working, in 1998, on the following standards with the provisional titles:

- WG1: Cleanliness Classification by Airborne Particulate Cleanliness
- WG2: Biocontamination
- WG3: Metrology and Testing Methods
- WG4: Design, Construction and Start Up
- WG5: Cleanroom Operation
- WG6: Terms, Definitions and Units
- WG7: Enhanced Clean Devices
- WG8: Molecular Contamination

CLEANROOM STANDARDS—INFLUENCING FACTORS

Contaminants and Other Factors to be Considered

Since the first Federal Standard 209, the most studied contamination in cleanrooms and clean zones are airborne particles. Over the following years, many other contaminates, i.e. factors that can harm the product/process/operator, were considered. The importance of these factors is dependent on the specific cleanroom demands and a cleanroom designer must know which of these contaminants should be considered. Table 2.1 serves as a check list.

Federal Standard 209 and other national standards consider only the total number of airborne particles; nothing else. For microbiological contaminants, pharmaceutical standards (Guides to Good Manufacturing Practice) as well as the ISO biocontamination control standards (see later in the chapter) should be consulted.

Cleanliness—particulate contaminants

The most measured and discussed contaminants are particles. They can be very different: dead or alive, inert or toxic, small or big, white or black, solid or liquid, rough or glossy. There is also a very great size-span, from 1000μm to 0.001 μm, and they can have

Table 2.1. List of possible 'contaminants' in a cleanroom.

Particulate and gaseous contaminants
Cleanliness of the air and surfaces of the working place
(dead and living particles, inert and toxic particles)

Environmental factors of other types

Temperature of the air	Noise
Humidity of the air	Radiation
Light intensity	Ionization
Vibrations	Electrostatics, ESD
Electromagnetics, EMD	

widely differing morphology and surface properties, which complicates their counting. A good knowledge of existing test methods and their limitations, as well as an appreciation of proper sampling techniques at very low concentrations and statistical interpretation is necessary to properly determine the cleanroom class. This problem is discussed in Federal Standard 209E.

Over the last 20 years most cleanroom classes have been based on the measurement of dead, airborne, particulate matter of a size $\geqslant 0.5$ μm or $\geqslant 5$ μm. Federal Standard 209E considers five particle sizes: $\geqslant 0.1$ μm; $\geqslant 0.2$ μm; $\geqslant 0.3$ μm; $\geqslant 0.5$ μm and $\geqslant 5\ 0$ μm. Also considered are 'ultrafines', which are defined as <0.02 μm, and large or 'macro' particles which are defined as $\geqslant 5$ μm. This way of classifying is used also by CEN and ISO in their air classification work.

Cleanroom classes for microorganisms (pharmaceutical or bioclean classes) also include microbe-carrying particles as a reference. The determination of living microorganisms in the cleanroom environment is usually complex and time-consuming (1–3 days) and has certain limitations which influence the count. Today there is an IES RP 023 on microbial measurements and ISO/CEN have issued a draft standard on the measurement of microorganisms in air, on surfaces, in liquids, and on textiles.

The relationship between inanimate and living particles has not been sufficiently studied and this subject is discussed later in this chapter. Federal Standard 209E states that there is no stable relationship. The new European Union Guide to GMP gives class figures for inanimate and living particles but in different tables. These figures are given later in this chapter.

The hazardous character of dangerous chemical aerosols must be known and occupational aspects taken into account. Here Threshold Limit Values (TLVs) are used instead of Classes. Not only toxic products must be considered, but allergic products, hormones, etc. as they are harmful to the personnel and products being produced nearby.

Relationship Between Class and Design Materials

To design, build and maintain a cleanroom properly, we must know the relationship between layout, design, material, operators, etc. and the 'class'. Greater cleanliness calls for certain design concepts and better materials for the inner surfaces of a room, as well as increased cleaning frequency, better cleanroom clothing, better changing rooms, etc. Long experience has shown that a correlation exists, despite the fact that cleanliness often refers only to air cleanliness and the air supply volumes.

This relationship between materials used in the room and the operators with respect to cleanroom classes, is given or stressed only in some cleanroom standards. One example is the IES RP for cleanroom 'garment systems' (IES-RP 003), where the type of garment and frequency of changes for different cleanroom classes is given.

Cleanrooms and Classified Rooms

In the early 1960s, the term 'white room' was used to describe a room with a low level of dust and with controlled temperature, humidity, etc. The walls and ceiling were often painted white. The term 'cleanroom' was then adopted.

Over the years there have been many definitions of cleanrooms. The Federal

Standard 209E published in 1992 gives a short definition as follows: 'A room in which the concentration of airborne particles is controlled and which contains one or more clean zones.' This is an important design statement in that it highlights the fact that a cleanroom can have several clean zones and acknowledges the important new design concepts of mini- and microenvironments in a cleanroom, which were formally developed in the 1990s.

The ISO cleanroom classification standard 14644-1 defines both a cleanroom and a clean zone. More details are mentioned in the definition and it defines a cleanroom as: 'A room in which the number concentration of airborne particles is controlled and which is constructed and used in a manner to minimize the introduction, generation and retention of particles inside the room and in which other relevant parameters, e.g. temperature, humidity and pressure, are controlled as necessary.'

Cleanrooms can have 'one or more clean zones', protecting a critical area. The environment for a clean process must not necessarily be a cleanroom, but can often be a clean zone in a less clean room. These clean zones can be provided by clean benches, work stations, minienvironments and isolators, the standards of which should also be considered by the designer.

In the above definitions 'control' is the key word. The word 'clean' is a little misleading as the poorest cleanrooms may not be very clean, their cleanliness spanning seven magnitudes (FS 209E) or nine magnitudes (ISO 146644-1). In the less-clean rooms (as in the hydraulic industry) the cleanrooms are not very clean, but they are controlled.

Contaminants other than particles and microorganisms must be considered and controlled. Because of this, the term 'controlled room' would be better. The term 'cleanroom' has, however, been used for a long time and the terms 'cleanroom' and 'clean zone' are used as a general term in this chapter.

Class Conditions to be Considered When Designing to a Certain Class

When designing a cleanroom to a given class the designer must refer to a defined condition or mode, of which there are three (usually one referred to the 'operational condition'). Usually the designer will design a cleanroom to only one condition, but it may be necessary to refer to another condition. The conditions are:

1. When the cleanroom is just built and ready for the installation of equipment.
2. When all equipment is installed and ready to run but without the contaminating influence of people.
3. When the room is operational and both the equipment is running and personnel are working.

Major classification standards call these three conditions 'as built', at rest' and 'operational'. The agreed definition of these conditions is given in the new ISO standard and given below in the relevant section.

It is normal for the contractor who installs the ventilation system to be asked to provide a cleanroom in the'as built' condition. To ensure that this cleanroom will then achieve the correct classification in the 'operational' condition, it is normal to design and build the cleanroom at least one class better in the 'as built' condition. Provided that the dispersion of particles from equipment and personnel is not excessive, then the correct 'operational' conditions should be achieved.

CLEANROOM CLASSES

Cleanroom classification standards can be divided into the following subgroups given below:

Engineering Classes:	These originate from Federal Standard 209, and are based on inanimate particles in air.
Biocontamination (Pharmacy) Classes:	These were originally based on Federal Standard 209 but developed to include living microorganisms. These standards are required for hygenic or sterile production. Guides to Good Manufacturing Practice and the CEN/ISO biocontamination standards cover this field.
Containment Classes:	These are for areas where hazardous contaminants are used or can occur.

The Present Engineering Classes

These classes are used mainly in rooms where electronic and engineering items are manufactured. Most of these standards are based on one of the various editions of the Federal Standard 209. Some countries completely adopted FS 209, while others made their own national version, similar to FS 209. Some made minor changes of the classes to comply with the metric system, but all changed the denomination of the classes (see Table 2.2). Because of the different naming of the classes in different countries, care must be taken not to mix up the standards.

The following major national '209-classes' exist today:

Australia	AS 1386	1989	Cleanrooms and clean work-stations
France	AFNOR X44101	1981	Definition of cleanroom levels
Germany	VDI 2083:3	1993	Contamination control measuring technique for clean air rooms
Holland	VCCN 1	1992	Dust and microorganism classification of air
Japan	JIS-B-9920	1989	Measuring methods for airborne particles in clean room and evaluating methods, etc.
Russia	Gost-R 50766	1995	Cleanroom classification, General requirements.
UK	BS 5295	1989	Environmental cleanliness in enclosed spaces
US	FS 209 E	1992	Airborne particulate cleanliness classes in cleanrooms and clean zones

Shown in Table 2.2 is a comparison of the classes given in some of these various standards as well as the ISO/CEN standard. The ISO standards are not binding on ISO members who have a similar national standard. However, when they are adopted and published by CEN as EN standards (probably starting in 1998), similar national

TABLE 2.2. A comparison of major engineering cleanroom classes in the world.

USA 209 E 1992		ISO 14644-1 1997	Japan B 9920 1989	France X44101 1981	Germany VDI 2083 1990	UK BS 5295 1989	Australia AS 1386 1989
		ISO Class 1	1				
	M 1	ISO Class 2	2		0		
1	M 2	ISO Class 3	3		1	C	0.035
10	M 3	ISO Class 4	4		2	D	0.35
100	M 4	ISO Class 5	5	4000	3	E, F	3.5
1000	M 5	ISO Class 6	6	—	4	G, H	35
10 000	M 6	ISO Class 7	7	400 000	5	J	350
100 000	M 7	ISO Class 8	8	4 000 000	6	K	3500
		ISO Class 9			7	L	

Note: the M-values for FS 209E (and the ISO-values) are in metric units. The M figures are therefore elevated above the line of the others which are given in cubic feet. FS 209E class 100 therefore corresponds to M class 3.5 and ISO 5.

standards within the EU countries will no longer be allowed. Therefore these new standards will be more binding within the EU than in the rest of the world.

It will take some time until all the ISO standards are ready and in use. It is therefore useful to consider existing national standards.

Federal Standard 209E, and its Four Early Editions   Federal Standard 209 was published in 1963, revised in 1966 (209A), in 1973 (209B), in 1987 (209C), in 1988 (209D) and in 1992 (209E). FS 209B was a mixture of information, advice and rules. It was very useful for its time, helping people to enter the new field of contamination control and building cleanrooms. Only seven pages were mandatory, while the 26 pages of its appendix were 'Non-mandatory guidance information'. The revision of version B took 14 years. Version C was printed in 1987 with printing errors and a corrected version D was therefore published within a year. Given in Table 2.3 are the cleanroom classifications found in FS 209 prior to version 209E. Despite this classification being superseded, it is still internationally used and the most common of all the methods used to

TABLE 2.3. Federal Standard 209 (A to D) class limits.

Class	Measured particle size (μm)				
	0.1	0.2	0.3	0.5	5.0
1	35	7.5	3	1	NA
10	350	75	30	10	NA
100	NA	750	300	100	NA
1000	NA	NA	NA	1000	7
10 000	NA	NA	NA	10 000	70
100 000	NA	NA	NA	100 000	700

NA = Not Applicable

describe the cleanliness of cleanrooms. This is the nomenclature method used in this book, although the equivalent ISO classification is also given.

Federal Standard 209—version E   A much changed version E was published in 1992, with a more precise title: 'Airborne particulate cleanliness classes for cleanrooms and clean zones'. Version E differs from the previous editions in that:

- the cleanroom classes are metric;
- it has seven classes of cleanliness: M1–M7;
- it gives a method for measuring air cleanliness;
- it demands a plan for monitoring air cleanliness;
- it gives a rational for the statistical rules used;
- ultrafine particles are considered, i.e. particles < 0.02 µm;
- it considers iso- and anisokinetic sampling;
- it describes sequential sampling for low concentrations of particles.

Federal Standard 209E has a table, which is reproduced here (Table 2.4), giving the class limits of the cleanroom in terms of the particle concentration in both metric and the original English units. This table does not necessarily represent the size distribution to be found in any particular situation.

Concentration limits can be calculated for intermediate classes, approximately, from the equation:

$$\text{particles/m}^3 = 10^M (0.5/d)^{2.2}$$

where M is the numerical designation of the class based on SI units and d is the particle size in micrometres.

British Standard: BS 5295   This standard, which is entitled 'Environmental cleanliness in enclosed spaces', was first published in 1976 and revised and published in 1989. It is similar to FS 209B in that it gives useful information to those designing and using a cleanroom. It is divided into five sections:

Part 0: General introduction, terms and definitions for clean rooms and clean air devices.
Part 1: Specification for clean rooms and clean air devices
Part 2: Methods for specifying the design, construction and commissioning of clean rooms and clean air devices
Part 3: Guide to operational procedures and disciplines applicable to clean rooms and clean air devices
Part 4: Specification for monitoring clean rooms and clean air devices

Shown in Table 2.5 is the classification table. When the new ISO classification is accepted by CEN the above classification will become obsolete. BS 5295 has been of value to the cleanroom designer because it considers details of the cleanroom and

TABLE 2.4. Federal Standard 209E airborne particle cleanliness classes.

Class name		0.1 µm Volume units		0.2 µm Volume units		0.3 µm Volume units		0.5 µm Volume units		5 µm Volume units	
SI	English	(m³)	(ft³)	(m³)	(ft³)	(m³)	(ft³)	(m³)	(ft³)	(m³)	(ft³)
M1		350	9.91	75.7	2.14	30.9	0.875	10.0	0.283	—	—
M1.5	1	1 240	35.0	265	7.50	106	3.00	35.3	1.00	—	—
M2		3 500	99.1	757	21.4	309	8.75	100	2.83	—	—
M2.5	10	12 400	350	2 650	75.0	1 060	30.0	353	10.0	—	—
M3		35 000	991	7 570	214	3 090	87.5	1 000	28.3	—	—
M3.5	100	—	—	26 500	750	10 600	300	3 530	100	—	—
M4		—	—	75 700	2 140	30 900	875	10 000	283	—	—
M4.5	1000	—	—	—	—	—	—	35 300	1 000	247	7.00
M5		—	—	—	—	—	—	100 000	2 830	618	17.5
M5.5	10 000	—	—	—	—	—	—	353 000	10 000	2 470	70.0
M6		—	—	—	—	—	—	1 000 000	28 300	6 180	175
M6.5	100 000	—	—	—	—	—	—	3 350 000	100 000	24 700	700
M7		—	—	—	—	—	—	10 000 000	283 000	61 800	1750

Class Limits

TABLE 2.5. BS 5295; Environmental cleanliness classes.

Class of environmental cleanliness	Maximum permitted number of particles per m³ (equal to, or greater than, stated size)				
	0.3 μm	0.5 μm	5 μm	10 μm	25 μm
C	100	35	0	NS	NS
D	1 000	350	0	NS	NS
E	10 000	3500	0	NS	NS
F	NS	3500	0	NS	NS
G	100 000	35 000	200	0	NS
H	NS	35 000	200	0	NS
J	NS	350 000	2000	450	0
K	NS	3 500 000	20 000	4500	500
L	NS	NS	200 000	45 000	5000
M	NS	NS	NS	450 000	50 000

NS = No specified limit

clean air devices. The United Kingdom has been a member of the CEN work since it started in 1990 and ISO work since its start in 1990. It is the convenor of the classification work and, since 1998, the Molecular Contamination work.

German Standard: VDI 2083 The German Engineering Association, known in Germany as Vereinigte Deutsche Ingenieure (VDI), has a group working in the field of contamination control. In 1976, based on FS 209, the VDI published VDI 2083 as their 'Cleanroom engineering' standard, with a metric system used in the classification. It was then revised from 1987 onwards and consists (in 1998) of eleven parts. These are:

Part 1. Fundamentals, definitions and determination
             of classes   (issued 1991)
Part 2. Construction, operation and maintenance   (issued 1991)
Part 3. Measuring technique for clean air rooms   (issued 1993)
Part 4. Surface cleanliness   (issued 1991)
Part 5. Criterion on comfort   (issued 1989).
Part 6. Personnel in cleanroom work area   (issued 1991)
Part 7. Cleanliness of process media (liquids, gases, etc.) (issued 1991)
Part 8. Suitability of products for cleanroom   (issued 1991)
Part 9. Quality, production and distribution of superpure
             water   (issued 1991)
Part 10. Media distribution   (issued 1998)
Part 11. Quality assurance   (under development)

VDI 2083: Part 3 gives the German cleanroom classification. It is a metric standard and uses a size of ≥1.0 μm as a basis of its nomenclature, instead of the ≥ 0.5 μm used by FS 209E.

The work on ISO contamination control standards is done in English, but when converted to CEN standards they are translated simultaneously into French and

German. Thus there are early ISO/CEN draft standards available in the German language covering many of the subjects shown in VDI 2083. Since 1996, a CEN pre-standard, the pr ENV 1631 about 'Design and construction' and, since 1998, a pr EN ISO 14644-1 have been available in the German language. Germany has been a member of the CEN work since its start in 1990, and the ISO work since its start in 1990. It is the convenor of the cleanroom design work.

Japanese Standard: Japanese Industrial Standard; JIS B 9920 Having been translated into Japanese, the Federal Standards 209 B to D were used by the Japanese as their standards until the late 1980s. A Japanese cleanroom classification standard JIS B 9920 was, however, published in 1989. It deviates from FS 209E in the following way:

- it is metric and has a denomination system based on the exponent of particle concentration;
- it uses particles $\geqslant 0.1\ \mu m$ as the reference size, instead of $\geqslant 0.5\ \mu m$ used in FS 209E;
- it has two cleaner classes than FS 209E.

However, it is fairly similar to the forthcoming ISO 14644-1, as the ISO cleanroom classification is also based on particles $\geqslant 0.1\ \mu m$, that method being derived from this standard. Japan has been a member of the ISO work since its start in 1993 and is also an observer at the similar CEN work. Japan is the convenor of the ISO work on metrology standards which will be used to measure the cleanroom classification.

Australian Standard: AS 1386 This was first published in 1976 and is now revised and was republished in 1989. It consists of seven parts. The Australians classify cleanrooms using a metric system. This is shown and compared with the other standards in Table 2.2 (see page 28). Australia has participated in the ISO work on cleanroom standards since 1994.

French Standard: AFNOR X 44101 In France, the French cleanroom association (ASPEC) has developed many 'Recommendations' for cleanrooms since 1972. Their main cleanroom classification was taken over and published in 1981 by the French organization for standards (AFNOR), and given the number X 44101. France has been a member of the CEN work since its start in 1990 and the ISO work since its start in 1990. France has therefore stopped work on national cleanroom standards. It was the convenor of the biocontamination work up to 1998.

Dutch Standard: VCCN-RL-1 In the Netherlands, contamination control 'Guidelines' have been issued by their cleanroom association, VCCN. They have issued six contamination control standards since then, the first being VCCN-RL-1 which considers the 'Particle and microorganism classification' of cleanrooms. VCCN-RL-2 is about 'Building and maintenance of cleanrooms'. The Netherlands has been a member of the CEN work since its start in 1990 and the ISO work. Despite this, their classification standard was published in 1992.

Russian Standard: GOST R 50766-95 In the 1970s the Soviet Union published a cleanroom classification standard with five classifications. It was based on the FS 209E. During the 1980s a similar COMECON standard was published in the Soviet Union.

In 1991 a Soviet Cleanroom Association (ASENMCO) was formed acting for international cooperation and standardization. Russia started to participate in the ISO work during 1993. The first Russian classification, GOST R 50766-95 'Cleanroom classifications. Methods of certification. General requirements', was developed by ASENMCO and approved by Gosstandard in January 1996.

The New ISO Classification Standard

This ISO standard 14644-1 is expected in 1999. The standard is divided into sections depending on whether the standard should be complied with (normative) or whether it provides information that may be used if required (informative). The standard contains the following sections:

Foreword: Explains that the document is one of a series of ISO cleanroom standards.

Specification (normative): Scope, definitions, classification. How to demonstrate that the cleanroom complies with the standard.

Annex A (informative): Definitions.

Annex B (informative): Graphical illustration of the cleanroom classes.

Annex C (normative): Method of testing for the determination of particulate cleanliness classification, using a discrete particle counter. It discusses the following:
(a) a discrete particle counter (DPC);
(b) pretest conditions;
(c) calculation of sampling location numbers;
(d) calculation of a single sample volume (minimum of three);
(e) procedure for sampling;
(f) recording of results;
(g) computing of 95% UCL;
(h) interpretation of results.

Annex D (normative): Procedure to calculate the 95% upper confidence limits of the particle counts.

Annex E (informative): Worked examples of classification calculations.

Annex F (informative): Counting and sizing of particles outside the size range considered for classification, i.e. ultrafine particles and macroparticles.

Annex G (normative): Sequential sampling procedure. This gives the following:
(a) background and limitations;

(b) basis for the procedure;
(c) procedures for sampling;
(d) figures for interpretation.

The normative section of the standard is thus the specification plus three appendixes. The normative specification deals with the following:

1. Occupancy states: Three states are defined in the mandatory part 2, Definitions. These are:

(a) As built: Condition where the installation is complete with all services connected and functioning but with no production equipment, materials, or personnel present.
(b) At rest: Condition where the installation is complete with equipment installed and operating in a manner agreed upon by the customer and supplier, but with no personnel present.
(c) Operational: Condition where the installation is functioning in the specified manner, with the specified number of personnel present and working in the manner agreed upon.

2. Classification number. This is based on the following formula:

$$C_n = 10^N \times \left[\frac{0.1}{D}\right]^{2.08}$$

where

C_n represents the maximum permitted concentration (in particles/m³ of air) of airborne particles that are equal to or larger than the considered particle size; C_n is rounded to the nearest whole number;

N is the ISO classification number, which shall not exceed the value of 9; intermediate ISO classification numbers may be specified, with 0.1 the smallest permitted increment of N;

D is the considered particle size in μm;

0.1   is a constant with a dimension of μm.

Table 2.6, published in the standard, provides an illustration of the classes that can be derived from the above formula:

3. Table 2.6 gives the possibility of a crossover from the new ISO classes to the old FS 209 classes. The new ISO classes can be converted by dividing the metric conversion factor, i.e. by 35. Thus the ISO 5 is equivalent to the old FS 209 Class 100. The ISO standard allows an unlimited number of classes, but nine main classes are given in the table.

4. Shown in Figure 2.1 is a graphical illustration of Table 1 as given in the standard (as in the FS 209E). The lines provide only an approximation of the class limits and may not be extrapolated beyond the lines, the lines not representing the actual particle size distribution.

TABLE 2.6. Selected ISO airborne particulate cleanliness classes for cleanrooms and clean zones.

Classification numbers (N)	Maximum concentration limits (particles/m³ of air) for particles equal to and larger than the considered sizes shown below					
	0.1 μm	0.2 μm	0.3 μm	0.5 μm	1 μm	5.0 μm
ISO Class 1	10	2				
ISO Class 2	100	24	10	4		
ISO Class 3	1000	237	102	35	8	
ISO Class 4	10 000	2370	1020	352	83	
ISO Class 5	100 000	23 700	10 200	3520	832	29
ISO Class 6	1 000 000	237 000	102 000	35 200	8320	293
ISO Class 7				352 000	83 200	2930
ISO Class 8				3 520 000	832 000	29 300
ISO Class 9				35 200 000	8 320 000	293 000

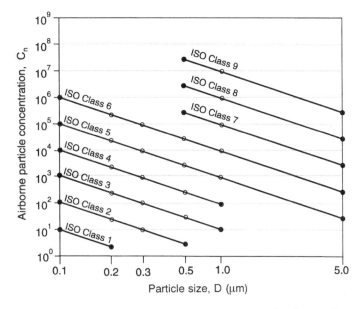

FIGURE 2.1. Graphical illustration of the airborne particulate cleanliness classes (Table 1 of ISO 14644-1).

5. Designation: It gives a method as how to express cleanliness classes, which should read as:

ISO Class N + occupancy state(s) applied + considered particle size(s)

When comparing the present Federal Standard 209E and the ISO Standard 14644-1 there are some important differences to be considered. These are shown in Table 2.7. Examples are given in both standards as to how to calculate the cleanroom class and advice is also given on sampling methods.

TABLE 2.7. Comparison between FS 209E and ISO Standard 14644-1.

Regarding	FED STD 209E	ISO 14644
1. Disposition	according to Federal regulation.	according to ISO
2. Particle range	0.1–10 μm	0.1–5 μm
3. Classification	according to *table, equation* given	according to *equation*
4. Formula for classes	$C=(0.5/D)^{2.2}\times10^{N}$	$C=(0.1/D)^{2.08}\times10^{N}$
5. Reference particle size	0.5 μm	0.1 = a constant
6. Number of classes	M 1 to M 7	ISO 1 to 9 (expandable)
7. Occupancy states or conditions	as built, at rest, operational	as built, at rest, operational
8. Occupancy state at test	not given	1 or more states.
9. Requirement of particle sizes for classification	1 or more	1 or more, where diameter of one is 1.5 × diameter of next smaller particle
10. Minimum sample volume	2.83 litres	2 litres
11. Minimum counts/ test	20	20
12. Sampling locations	according to three given equations	number of locations=$\sqrt{}$ area of room
13. Number of samples/location	1 or more	1 but 3 if one location
14. Measurement methods	DPC (=OPC, CNC) +optical microscopy	DPC for 0.1–5 μm particles, CNC for ultrafines; optical microscope for macroparticles
15. Ultrafine particles.	considers particles ≤ 0.02 μm	considers particles ≤0.1 μm
16. Macroparticles ≥5 μm	not considered	information on particles ≥5 μm
17. Confidence level needed	95% UCL for ≤9 sample locations	95% UCL for ≤9 sample locations
18. Sequential sampling	is included	is included
19. Isokinetic sampling	is included	is not included

THE BIOCONTAMINATION AND PHARMACEUTICAL CLASSES

Production of Sterile Pharmaceuticals

The development of cleanroom standards for the pharmaceutical industry started during the early 1960s to counteract contamination problems of sterile pharmaceutical products which had caused sickness and death in patients. It had been realized that as only a small sample of the drugs could be tested, a final test of sterility could never determine the safety of sterile products. It was necessary therefore to rely on proper manufacturing, i.e. good manufacturing practice (GMP). These standards are known as Guides to Good Manufacturing Practice (GGMP). They were based on experience of the then new FS 209, BS 5295 and other engineering standards, as well as experience in the manufacture of pharmaceutical products. The goal of the GGMPs is to carefully specify the proper method of manufacture of sterile products by eliminating microbial and particle contamination and hence creating the correct quality assurance (QA).

The GMP guides deal largely with the methods of good manufacturing but also specify the building design, building material, care of personnel, etc. They may also give figures for the cleanliness classes of cleanrooms: particles, microorganisms, as well as the type of air filters and number of air changes per hour.

The different GMP guides, produced by the various countries, are advisory rather than legal and consider that there is more than one way to achieve the recommended standards. To check that the GMP is being applied correctly, each country has government inspectors. They will interpret the generally expressed statements of the guides. A pharmaceutical manufacturer must also comply with the GMP guides of the countries receiving their products.

The GMP guides are intended for the pharmaceutical industry, but the methods used to solve QA problems are generally useful. Besides the recommendations mentioned above, the methods of qualifying the equipment and validating the processes are applicable to other industries like food and microelectronics. GMP guides should therefore be studied in other industries with quality problems.

The determination of 'critical zones' in the room is important and a key factor in risk analysis, such as Hazard Analysis of Critical Control Points (HACCP). This method is considered in the new ISO biocontamination standards.

In 1963, the first GMP was published by the Food and Drug Administration (FDA) in the United States. Three years later a Swedish regulation was made official and a further three years later, in 1969, the first guide from the World Health Organization (WHO) was published. An important guide was then published by the European Pharmaceutical Inspection Convention (PIC) in 1972, and revised in 1983, 1989, 1992 and 1995 (Document PH 5/92). This PIC GMP Guide is accepted in most Western European countries, in a few East European countries, and in Australia.

In 1989 a guide known as the 'EEC Guide to Good Manufacturing Practices for Medicinal Products' was produced by the Commission of the European Communities. Reprinted in 1992 with new annexes, it came into legal use and superseded all other GMP guides produced in EEC countries. This version is now revised with a new cleanroom classification method. The EU GMP Annex 1 was valid from 1 January 1997.

Early in 1992, Russia issued its first GMP, LD 64-125-91, as the first GMP issued in Eastern Europe.

The most used GMP guides for cleanrooms are the following:

- PIC: GMP and Guidelines 1995   Valid in European countries outside the EU and Australia.
- FDA cGMP, 1987   Valid for the United States.
- EU GGMP 1997   Valid for the EU area.

Unfortunately, these GMPs are not harmonized, as shown in Table 2.8. The conditions for testing (at rest, or operational) are different and the letters for the level, or grade (A to D), do not mean the same in different countries.

There was criticism of the PIC and 1992 EEC GMP guide during the 1990s and a revised version was therefore developed and issued for use in 1997. The revised classification scheme is given in Table 2.9.

TABLE 2.8. Comparison of major pharmacy GMP guides regarding working conditions and classes (classification given according to FS 209E).

GMP	PIC GMP	EU-GMP 1992	FDA 1987	Transfer of classes, all operational		
Condition	At rest	Operational	Operational	209D	209E	ISO
Grade A	100	100	Critical area	100·	M3.5	ISO 5
Grade B	100	10 000	—	100	M3.5	ISO 5
Grade C	10 000	100 000	—	10 000	M5.5	ISO 7
Grade D	100 000	—	Controlled area	100 000	M6.5	ISO 8

TABLE 2.9. EU GMP Guide 1997. Air particle classification system for the manufacture of sterile products.

Grade	Maximum permitted number of particles/m³ equal to or above			
	At rest (^{b})		In operation	
	0.5 µm	5 µm	0.5 µm	0.5 mµ
A	3500	0	3500	0
B(^{a})	3500	0	350 000	2000
C(^{a})	350 000	2000	3 500 000	20 000
D(^{a})	3 500 000	20 000	not defined (^{c})	not defined (^{c})

(^{a}) In order to reach the B, C and D air grades, the number of air changes should be related to the size of the room and the equipment and personnel present in the room. The air system should be provided with appropriate filters such as HEPA for grades A, B and C.
(^{b}) At rest should be received in the unmanned state after the 15–20 min 'clean up' period.
(^{c}) Appropriate alert and action limits should be set for the results of particulate and microbiological monitoring. If these limits are exceeded, operating procedures should prescribe corrective action.

TABLE 2.10 The EU GMP Guide (1997) values for microbiological monitoring in the operational state, for the manufacture of sterile products.

Grade	Recommended limits for microbial contamination (^{a})			
	Air sample cfu/m³	Settle plates (diam. 90 mm), cfu/4 hours(^{b})	Contact plates (diam. 55 mm) cfu/plate	Glove print, 5 fingers cfu/glove
A	< 1	< 1	< 1	< 1
B	10	5	5	5
C	100	50	25	—
D	200	100	50	—

^{a}These are average values.
^{b}Individual settle plates may be exposed for less than 4 hours.

The maximum concentration of microorganisms that should be found in the various grades of cleanrooms in the operational state are given in Table 2.10

The EU GMP Guide gives no information on how to measure microorganisms. Test methods will be found in the forthcoming ISO biocontamination control standards and the IEST RP 023 (see below).

Manufacturing operations in the EU GMP guide are divided into two categories: those where the preparation is sealed in its final container and terminally sterilized, and those that must be prepared in an aseptic way at some or all stages. The guide suggests the following use of cleanrooms:

Terminally sterilized products

Grade	Examples of operation
A	Filling of products, when unusually at risk
C	Preparation of solutions, when unusually at risk. Filling of products.
D	Preparation of solutions and components for subsequent filling

Remarks: Preparation of most products should be done in at least a grade D environment, where there is an unusual risk, grade C environment should be used.

Aseptic preparations

Grade	Examples of operation
A	Aseptic preparation and filling
C	Preparation of solutions to be filtered
D	Handling of components after washing

Remarks: After washing, components should be handled in at least a grade D environment. Handling of sterile starting material should normally (see GGMP) be done in a grade A environment with a grade B background. The preparation of solutions which are to be sterile filtered during the process should be done in a grade C environment; if not filtered, the preparation of materials and products should be done in a grade A environment with a grade B background. The handling and filling of aseptically prepared products should be done in a grade A environment with a grade B background. The preparation and filling of sterile ointments, creams, suspensions and emulsions should be done in a grade A environment, with a grade B background, when the product is exposed and is not subsequently filtered.

Special conditions are specified in the GGMP when isolators or blow–fill–seal machines are used.

Other Biocontamination Class Standards

Methods to count airborne microbes were developed in the 1940s and automatic optical counting of dust particles in the 1960s. Microbial methods are, however, very time-consuming compared with optical or discrete particle counters (OPC or DPC) and require skilled technicians. It would therefore be convenient if a relationship between dead and living particles in the air could be established, then the fast OPC methods could be used for rapid indications. It is known that for a given room, parallel tests can indicate an approximate relationship, but there are many factors that must be considered and figures must not be transferred to a different cleanroom or factory.

In 1993 the IEST in the United States issued a RP 023 on 'Microorganisms in cleanrooms'. It gives methods and describes devices to measure microorganisms in air and on surfaces but it does not discuss cleanroom classification. It also deals with disinfectants and their influence on the measurements of microbial counts.

In the ISO 14698 standard on 'Biocontamination', methods are given for measuring microorganisms in air and liquids, and on surfaces, including fabrics.

ISO biocontamination standards: 14698 These comprehensive ISO biocontamination standards, which were in draft form at the time of preparation of this chapter, discuss general microbial classification and methods for microbiological testing. They are different from the pharmaceutical GGMPs because of their general application to the food industry, hospitals, the cosmetics industry, etc. They also have a different concept, as they do not give microbial class figures but require a local risk analysis (e.g. Hazard Analysis of Critical Control Points, HACCP), and risk classes. It also stresses the importance of determining values for target, alert and action levels.

The ISO standard also gives information on methods for the determination of microorganisms in air, liquids, and on surfaces (including textiles). It has three major parts:

ISO 14698-1: Biocontamination Control; General principles
ISO 14698-2: Evaluation and interpretation of biocontamination data
ISO 14698-3: Biocontamination control of surfaces

THE CONTAINMENT CLASSES

The handling of toxic or pathogenic material, as well as genetic engineering work, must be performed in special premises using containment equipment. Premises, equipment, and working techniques are regulated by these 'Standards'. In the United States guides on biosafety, which have been developed by the National Institute of Health, are contained in 'Recombinant DNA research—actions under the Guidelines: Federal Register 56' (1991). In Europe, the designer must meet the standards given in the EC Directive 90/219/EEC of 1990 when designing facilities for genetically modified organisms. A European standard 'EN 1620: Biotechnology—Large-scale process and production—Plant building according to degree of hazard', was issued in 1997. Further information on containment standards is given in Chapter 6 of this book.

Biological risks, and the measures are directed toward these, are classified in four classes. Four laboratories with risk levels of BL 1 to BL 4, are used (see Table 2.11).

In 1993 the ICCCS surveyed the 'Safety facility standards and practices', which gave 42 references but also showed the need for the international harmonization of standards in the biohazards field.

To be found in containment rooms is the safety cabinet, of which there are three types: Classes I, II and III. Class III gives the highest protection. Table 2.12 gives examples of Class II safety cabinet standards. A European standard should be issued

TABLE 2.11. The four classes BL 1–BL 4 of biosafety laboratories.

Class	Explanation	Example
BL 1	Normal laboratory standard	Ordinary biochemistry labs, school and university labs
BL 2	Special training and routines to prevent lab infections. Appropriate waste handling	Diagnostic labs Health labs
BL 3	Special lab with negative pressure Air locks for people and material Autoclave in the room All work done in safety cabinet Special decontamination of waste	Special safety labs Tuberculosis labs
BL 4	Special labs with total separation between humans and microorganisms in every respect, negative pressure, sterilization	High risk labs

TABLE 2.12. Safety cabinet (usually Class II) standards.

Standard number	Year	Country	Name of standard
NSF No 49,	1976	US	Class II biohazard cabinetry
BS 5726	1992	UK	Specification for microbiological safety cabinets
DIN 12950	1984	D	Sicherheitswerkbänke (in German)
JACA No. 16, 17	1980	J	Standard of biological safety cabinet
AS 2252.1-2	1989	AUS	Biological safety cabinets, Classes I and II
R^3-NORDIC	1988	Nordic R^3	Norm for safety benches (in Swedish)

shortly. This is EN 12469, entitled 'Biotechnology—Performance criteria for micro-biological safety cabinets'.

OTHER STANDARDS FOR THE CLEANROOM

As mentioned in the introduction, much work is going on to create international clean-room standards. These are standards for the specification of the design of cleanrooms, standards for monitoring that a cleanroom fulfils the class demands to which it was designed, standards to control the designed surface cleanliness, etc. There are also several Recommended Practices of the IEST that are of value to the designer. Some of these are described below.

Cleanroom Design Standards

There are a number of cleanroom design standards available which should be con-sidered. These are given in Table 2.13. These give the following information:

TABLE 2.13. Cleanroom design standards.

1. Consideration in cleanroom design	IEST RP 012 (for the US)	1993
2. Design, construction and operation of cleanroom and clean air devices	CEN prENV 1631 (for Europe)	1996
3. Design, construction and start-up of cleanroom facilities	ISO 14644-4 and EN 14644-4: (International and European)	expected 1999
4. GMP Guide (advice on design, etc.)	EU GMP (for Europe)	1997

1. Consideration in cleanroom design—IEST RP 012, issued 1993.
The IEST has developed a Recommended Practice that considers the performance criteria to be considered in the design of cleanroom facilities. It covers the following major topics:

- *Planning procedures.* This part gives a mission statement and then considers the goals and strategy with regards to critical material, flow, special requirements, etc. It also considers the determination of needs.
- *Design requirements.* The following are considered: airflow pattern, air velocity, HVAC, temperature, humidity, pressure, exhaust, submicron contamination, microorganisms, noise, vibration, electrostatic interference, grounding systems, cleanroom lighting, locally controlled environments: micro- and minienvironments.

There are two informative appendices:

Appendix A: Industry codes and standards, etc.
Appendix B: Reviewing and testing conditions

2. Design, construction and operation of cleanroom and clean air devices: CEN prENV 1631, 1996.
This is a European pre-standard, based on the work carried out between 1990 and 1993 by CEN on cleanroom design. It has a likely lifetime of two years and will be superseded by the extended new design standard, ISO 14644-4 (see below) and EN 14644-4 for Europe.

3. Design, construction and start-up of cleanroom facilities—ISO 14644-4, expected to be issued in 1999
This future ISO standard is based on a CEN draft from 1990-93, and is expected to be issued as an official standards by both ISO and CEN. It is a standard for general cleanroom facilities and is valuable because it is not process-specific and can be used in all areas where various quality of cleanrooms are used.

It specifies the requirements for the design and construction of cleanrooms and clean air devices (both called clean installations) but does not prescribe specific technological or contractual means to meet the requirements. It is intended for use by purchasers, suppliers and designers and provides a checklist of important parameters of performance.

Constructional guidance is provided, including requirements for start-up and qualification. Basic elements of design and construction that are needed to ensure continued satisfactory operation are identified through the consideration of relevant aspects of operation and maintenance.

It does not consider specific processes to be accommodated in the cleanroom. Only cleanroom construction specific requirements of initial operation and maintenance are considered. It gives normative lists on:

- Specification of requirements;
- Planning and design;
- Construction and start-up;
- Testing and approval;
- Documentation.

Documentation says that 'Details of a complete installation, all operation and maintenance procedures shall be documented. Documents shall be made readily available to all personnel responsible for start-up, operation and maintenance of the installation'.

The many annexes give information on cleanliness concepts with classification examples, development and approval of an installation, layout, construction, materials, air filtration, further specification and a comprehensive checklist. Finally, a bibliography is supplied.

The normative section on *Specification of requirements* is an important document for a proper and necessary agreement between supplier and purchaser. Such agreements are easily ignored if not required in this way.

4. Cleanroom design in EU GGMP, 1997.
Annex 1 of this GGMP defines cleanroom classes, different technologies for production, requirements for premises and equipment, and it elaborates on design questions. When designing a pharmaceutical cleanroom, it should be consulted.

Isolator and Minienvironment Design

The use of isolators and minienvironments is relatively new in cleanrooms. There is therefore a limited availability of design guidance. In the United States a working group of the IEST has been working since 1993 on a Recommended Practice for 'Minienvironments'. Since 1996 there has been an ISO working group (no. 7) considering 'Minienvironments and isolators', a title that was changed in 1998 to 'Enhanced clean devices'. It presented in 1998 a draft 14644-7 on 'Enhanced clean devices'. This gives useful design suggestions and references.

In the United Kingdom a guide was published in 1995: *Isolators for Pharmaceutical Applications.* HMSO, ISBN 0 11 701829 5.

Recommended Practices (RPs) of the Institute of Environmental Sciences and Technology (IEST), USA

The United States has, up to now, produced the greatest number of cleanroom standards. These started in the 1950s with the many military and space standards. Since 1980, the Institute of Environmental Science and Technology has taken a lead and created about 30 working groups to write cleanroom standards, including the revision of FS 209. In 1997 they had created about 25 such RPs and work is going on to extend and update them.

The RPs of IEST are not 'legal' standards but 'Recommended Practices', to assist those in the cleanroom business. They will, however, gradually be made official by ANSI, the standards organization of the United States.

The denomination of the RPs has a figure added which denotes the edition. RP 004.1 means a first edition of the RP4 and RP 004.2 means a second, usually a revision. Existing RPs, which are likely to have an impact on the cleanroom design work, are given below in Table 2.14. No indication is given as to what is the current edition is, since it is assumed that the most recent edition will be sought.

IEST has compiled all their Recommended Practices, including FS 209E and Military Standard 1246C, into an *IEST Handbook of Recommended Practices*.

TABLE 2.14. IEST Recommended Practices of value to a designer.

RP 001	HEPA and ULPA filters
RP 002	Laminar-flow clean-air devices
RP 003	Garment system considerations for cleanrooms and other controlled environments
RP 006	Testing cleanrooms
RP 009	Compendium of standards, practices, methods, and similar documents relating to contamination control
RP 012	Considerations in cleanroom design (see also later in this chapter).
RP 015	Cleanroom production and support equipment
RP 018	Cleanroom housekeeping—operation and monitoring procedures
RP 021	Testing HEPA and ULPA filters
RP 022	Electrostatic charge in cleanrooms and other controlled environments
RP 024	Measuring and reporting vibration in microelectronic facilities
RP 028	Minienvironments

Standards for Surface Cleanliness

Cleanroom classification has, until now, been almost entirely on the basis of particles or living microorganisms in the air. In practice, the influence of these contaminants often occurs because of their deposition onto or into material. Methods have been developed to study the fallout or impact of particles on surfaces. In the 1997 EU GMP Guide, settling figures for living microbiological particles in different cleanroom classes are given. In the aerospace and electronic industry, so-called 'settling' or 'witness' plates have been used to evaluate the quality of surfaces in the work place. In both Germany (VDI) and in the United States (IEST) work has been ongoing to define surface cleanliness classes. This work is to be found in VDI 2083 part 4 and Mil. Std. 1246.

The designer should consult the buyer of the cleanroom if they specify sedimentation values, as well as the ordinary ISO class for airborne particles.

CLEANROOM STANDARDS FOR DESIGN PURPOSES, TO SELECT TECHNICAL CONCEPTS AND SOLUTIONS, MATERIALS, EQUIPMENT, ETC.

A standard often contains two types of information; that which must be carried out (normative) and that which is useful advice (informative) and is often contained in annexes. Often the advisory part is much larger than the normative. Because standards are generally not written for a specific cleanroom industry and must not bind technical development, the way they are written must be general. They will therefore not give engineering details that are specific for a pharmacy, electronics, etc. and to those looking for specific practical details and solutions, this is a disappointment. However, by reading the text and consulting experienced people, practical, detailed solutions can be found.

As an example of this method of writing, advice on airlocks can be found in BS 5295 and in the EU GGMP. The advice given in these two standards is helpful but not specific. We read for *airlocks* that:

BS 5295 (1989) Part 2: A.2.2 Guidance

> In order to maintain overpressure and integrity of the controlled space during entry and exit at least one air lock will normally be required.
>
> Barrier benches or other demarcation systems and decontamination devices and procedures should be employed within an airlock system for the passage of material. Precautions should be taken to ensure that entry and exit doors associated with an airlock are not opened simultaneously. In both cases doors should be provided with clear windows such that a line of sight view is provided between them. Consideration should be given to the use of electric and/or mechanical interlock system including audio-visual indicators.
>
> Barrier benches or other demarcation systems and decontamination devices and procedures should be employed within an airlock system for the passage of material. Consideration should be given to segregation of passage of material and personnel.

EU GMP (1997) Annex I:28

> Both airlock doors should not be opened simultaneously. An interlocking system or a visual and/or audible warning system should be operated to prevent the opening of more than one door at a time.

Table 2.15 gives a number of valuable standards which are useful to consult when considering the design requirements for materials, equipment, etc. to be used in a cleanroom. Another way to find standards and other reference material is to visit the Home Page of the International Confederation of Contamination Control Societies on the Internet. This is found on: `http//www.icccs.org.`

CLEANROOM STANDARDS

Table 2.16 gives a list of relevant cleanroom standards, divided into the three major fields of Engineering, Pharmacy, and Containment. The name of the publisher is given in abbreviated form but their full name and address is given in the next section of the chapter.

TABLE 2.15. Valuable design standards.

FDA	FDA, USA; 'Guide to GMP'	1987
BS 5295	British Standard BS 5295, Parts 1 and 2	1989
FS 209 E	Federal Standard 209E (mainly the Appendices)	1992
PIC GMP	PIC 'Guide to GMP', PH 5/92	1992
IES RP 012.1	IES Recommended Practice: 'Considerations in cleanroom design'	1993
HMSO -95	Isolators for Pharmaceutical Applications, ISBN 0 11 701829 5	1995
EU GMP	EU-GMP: 'Guidelines for the manufacture of sterile medicinal products'. Annex 1 of GMP	1997
ISO 14644-1	ISO Draft on 'Classification of airborne particulates', part of its annexes	completed in 1999
ISO 14644-4	ISO Draft on 'Design, construction and start-up of cleanroom facilities'	near completion in 1999

TABLE 2.16. Relevant cleanroom design standards

Designation and country	Year		Language	Published by
Engineering standards:				
Airborne Particulate Cleanliness Classes in Cleanrooms and Clean Zones				
USA	1992	FS 209E	English	GSA
Consideration in Cleanroom Design				
USA	1993	IES-RP-CC 012.1	English	IEST
Cleanrooms and Work-Stations—Principles of Clean Space Control				
Australia	1989	AS 1386.1	English	SAA
Environmental Cleanliness in Enclosed Spaces				
Britain	1989	BS 5295	English	BSI
Définition et Classification de la Propreté Particulaire de l'Air et d'autres Gaz				
France	1981	X44101	French	AFNOR
Clean Room Engineering—Fundamentals, Definitions, and Determination of Cleanliness Classes				
Germany	1990	VDI 2083, part 1	German & English	VDI
Clean Room Engineering—Construction, Operation, Maintenance				
Germany	1992	VDI 2083, part 2	German & English	VDI
Measuring Methods for Airborne Particles in Clean Rooms and Evaluating Methods for Air Cleanliness of Clean Rooms				
Japan	1989	JIS B 9920	Japanese, English translation available from JIS	
Classification of Airborne Particulate Cleanliness for Cleanrooms and Clean Zones				
ISO	(1998)	ISO 14644-1	English	ISO
Specification for Testing and Monitoring Cleanrooms and Clean Zones to Prove Continued Compliance with ISO 14644-1				
ISO	(1998-9)	ISO 14644-2	English	ISO

TABLE 2.16 *Continued*

Designation and country	Year		Language	Published by
Pharmaceutical standards:				
GMP for Human and Veterinary Drugs				
USA	1976	21 CFR Part 211	English	FDA
Validation of Aseptic Filling for Solution Drug Products				
USA	1980	Techn Monogr No 2	English	PDA
Guide to GMP for Medicinal Products				
EC	1992		English	EEC
Guide to GMP for Pharmaceutical Products.				PIC at EFTA
Europe	1992	PH 5/92	English/German/ French Secretariate	
Guidelines on Sterile Drug Products, Produced by Aseptic Processing				
USA	1987		English	FDA
Biocontamination Control (draft)				
ISO	1999	ISO DIS 14698-1, 2, 3	English	ISO
Containment standards:				
EC Directive 90/219/EEC Europe	1990		English	EU, Brussels
Federal Register 56: Recombinant DNA Research—Action and Guidelines				
USA	1991		English	NIH
EN 1620: Biotechnology: Large Scale Processing and Production	1997		English	EuroNorm, Brussels

ISO standards like ISO 14644-1 are available from ISO in Geneva or from national standards organisations or institutes (like the BSI in the United Kingdom). CEN documents can only be purchased from national standards organisations or institutes. Pharmaceutical documents are usually available from governmental pharmacy boards in each country or from international organisations like EPC or WHO whose addresses are given below.

ABBREVIATIONS/SOURCE CODE

The following list gives both an explaination of the abbreviations given in the text as well as sources of information. As well as the sources given in this list there are 16 contamination control societies who are useful sources of information. These societies, who are members of ICCCS, represent Australia, Belgium, Brazil, China, Denmark, Finland, France, Germany, Italy, Japan. The Netherlands, Norway, Russia, Scotland, South Korea, Sweden, Switzerland, the United Kingdom and the United States. Their Secretariates can be reached from information available on the ICCCS web site: http//www.icccs.org.

ANSI	American National Standards Institute, 1430 Broadway, New York, NY 10018, USA.
ASHRAE	American Society of Heating, Refrigeration and Air-Conditioning, 1791 Tullie Circle NE, Atlanta, GA 30329, USA.
ASTM	American Society for Testing and Materials, 1916 Race Street, Philadelphia, PA 19103, USA.
BSI	British Standards Institution, 389 Chiswick High Road, London W4 4AL, UK.
CEN	Committee European Normalization, Rue De Stassart 36, 1050 Brussels, Belgium.
DFG	Deutche Forschungsgemeinschaft, Kennedy Allé, D-5300, Bonn 2, Germany.
DIN	Deutsches Institute für Normierung e.V. Normenausschuss Medicin, Beuth Verlag, GmbH, D-1000 Berlin 30, Germany.
EPC	European Pharmacopia Commission, Strassbourg, France.
EUR-OP	Office for Official Publications of the European Community, 2 Rue Merciere, 2985, Luxemburg.
EUROVENT	Secretariate of EUROVENT, (European Committee of HVAC Equipment Manufacturers) Bauernmarkt 13, Vienna, Austria.
FDA	Food and Drug Administration Division of Drug Quality Compliance, Center for Drugs and Biologics, 5600 Fishers Lane, Rockville, MD 20857, USA.
FIP	Federation International Pharmacie, The Haag, The Netherlands.
GSA	General Services Administration, GSA Service Center, Seventh & D Streets, SW, Washington, DC 20407, USA.

HIMA	Health Industry Manufacturers Association, 1030 15th Street NW, Washington DC, USA.
HMSO	Her Majesty's Stationery Office, 49 High Holborn, London WC1V, UK.
ICCCS	International Committee of Contamination Control Societies, c/o SRRT, Switzerland (see below).
IEST	Institute of Environmental Science and Technology, Mt Prospect, IL, 60056–3444, USA.
ISO	International Organization for Standardization, TC 209 Alexander Gorshkov, 1 Rue De Varembe, CH-1211 Geneve 20, Switzerland.
IKH	Interkantonale Kontrollstelle für Heilmittel. (Intercantonale Office for the Control of Medicaments, IOCM) 3000 Bern 9, Switzerland.
JIS	Japanese Industrial Standards Committee, Standards Department, Ministry of International Trade and Industry, 1–3–1 Kasumegaseki, Chiyoda-ku, Tokyo 100-8921, Japan
NCI	National Cancer Institute, Bethesda, MD 20205, USA
NIH	National Institute of Health, Bethesda, MD 20205, USA.
NSF	National Sanitation Foundation 3465 Plymouth Rd, P.O. Box 1468 Ann Arbor, MI 48106, USA.
PIC	Secretariate of PIC, 9-11 Rue de Varembé, Geneva 20, Switzerland
SAA	Standards Association of Australia, Standards House, 80 Arthur St., North Sydney, NSW-2059, Australia
SRRT	Schweizerische Gesellschaft für Rein Raum Technik, Langwisstrasse 5, CH-8126 Zumikon, Switzerland.

VDI	VDI Verlag Publication (publications) PO Box 1139, D-4000 Düsseldorf, Germany.
VDI	Verein Deutsche Ingenieure, Graf-Recke Straße 84, D-4000 Düsseldorf, Germany.
WHO	World Health Organization, Geneva, Switzerland.

ACKNOWLEDGEMENT

Table 2.5 is reproduced by permission of the British Standards Institution.

3 The Design of Cleanrooms for the Microelectronics Industry

J. G. KING

INTRODUCTION

The technology of cleanroom design had its conception in medical/pharmaceutical facility development, and it was born in the rush to expand World War II's science of armament and warfare. Its adolescence paced the needs of the race into space and finally came into adulthood in time to meet the challenge of manufacturing in the Lilliputian world of microelectronics, where micrometre-sized particles are like automobile-sized rocks, and impurities of the order of parts per trillion can be crucial.

The demands of creating an environment which provides the near-perfect conditions required in the manufacturing processes of this world of microelectronics were, and are, stringent and include areas which are vital to the success of manufacturing. Maximizing product quality and yield requires strict control of, and co-ordination between, the manufacturing facility, tooling, process and operation, but all of these heavily depend upon the strict environmental control afforded by advanced cleanroom technology.

Contamination control engineering, along with its related equipment, materials, and skills, has become so advanced that the facility and its environmental control should no longer need be the limiting factor. If this is so, where is the challenge in designing cleanrooms for microelectronics manufacturing? Obviously, the goal of cleanroom environmental control is to provide contamination-free space in which to manufacture a contamination-free product. Following the approved procedures should produce the desired result, but contamination has a way of occurring unexpectedly, sometimes without any sign to indicate its origin. The cleanroom, the manufacturing process, equipment and method of operation are all equally possible sources of contamination. While the technology was yet in its youth, the immediate assumption was, and generally justifiably so, that the cleanroom itself had failed. Present-day cleanrooms are of a much higher standard and have controls and monitors by which the cleanroom's operating condition can be quickly evaluated. If it proves to be operating properly, the process, process equipment and operation can then be examined in turn.

The world of environmental control has truly expanded, because, to be recognized as a contamination-control engineer, one must now accept the challenge of including domain and responsibility. One must also be involved in evaluating handling equip-

Cleanroom Design. Edited by W. Whyte © 1999 John Wiley & Sons Ltd

ment, and training personnel in contamination control, as well as all the other innumerable facets of the total control problem.

MANUFACTURING SEMICONDUCTOR CIRCUITS

Manufacturers of semiconductor circuits have become very large users of cleanrooms. The reason is that the manufacturing operations take place at almost the molecular level and the physics of the operation of the device depend upon purity of materials in atomic percentages measured down to parts per trillion. These levels are unheard of in any other human endeavour.

First germanium and then silicon became the wonder elements of the twentieth century. This is because they are semiconductors of electricity. Semiconductors derive their name from their ability sometimes to conduct an electrical current and other times not. This is controlled by their internal structure on the atomic level and by the circuitry into which they are inserted. Certain atomic elements can be added to the structure during manufacturing which enables these effects to take place. While there are many described categories of devices (taken on a discrete device level), the basic functions performed are: voltage amplification, switching, resistivity and induction. Combined with diodes and transistors these form the basis of semiconductor circuits.

To give some perspective to those who are unfamiliar with the process, the following is a brief summary, together with levels of contamination control required, of the manufacturing steps used in making semiconductor circuits. The very-large-scale-integration (VLSI) and the complementary-metal-oxide-semiconductor (CMOS) processes on silicon are used as an illustrative example.

Common practice divides the manufacturing of integrated circuits into three phases: materials, wafer fabrication and assembly and test.

Materials

Starting with silica sand, the process is as shown in Table 3.1.

During the sequence most of the operations take place in an ordinary factory environment. Protection against contamination is provided, for the most part, by doing the processing within sealed systems. Preparation of the charge for the Czochralski puller, however, as well as clean-up of the crucible, must be done in a well-controlled cleanroom (minimum standard ISO 5 (Class 100)). Any contamination introduced during that step will get into the ingot and will either cause a failure in the process to make a single crystal or cause unacceptable electrical properties to develop.

The last part of the process also requires a clean environment (minimum standard is ISO 5 (Class 100)) to allow the equipment to produce the required finish, thickness, taper and flatness. Also, great care must be taken to avoid leaving any mobile ions or doping elements on the surface. Subsequent high temperature operations could then distribute these into the crystalline structure and destroy the desired electrical properties.

This is a vastly simplified description of the entire process, completely ignoring the many tests, cleaning steps and process adjustments required to produce material of the required purity, crystalline size, orientation, uniformity and possessing the desired

TABLE 3.1

Start with	Into	Produces
a. Sand, coal, coke, wood chips	Submerged arc electric furnace @ 2000°C	Silicon metallurgical grade (MGS)—98% Si
b. MGS (powdered) HCl (gas) + catalyst	Reactor @ 300°C	$SiHCl_3$—impure (trichlorosilane)
c. $SiHCl_3$ (impure)	Distillation column	$SiHCl_3$—pure
d. $SiHCl_3 + H_2$	Reactor	Silicon, electronic grade (EGS) 99.99% +
e. EGS	Czochralski crystal grower— high temp. inert gas atmosphere	Single-crystal silicon ingot
f. Single crystal silicon ingot	Special grinder	Polished cylinder with flat(s)— full length—to define crystal orientation
g. Polished cylinder	Diamond saw	Round slices with one or two flat sides
h. Slices	Grind, lap and polish machines	Wafers ready to be made into integrated circuits

electrical properties. The description is also based on using only silicon as the semi-conducting material. There are numerous other systems in use today, involving multi-metal semiconductors, silicon (or other semiconductors) deposited on glass, sapphire, diamond, etc. These systems do not negate any requirement for right control of contamination. On the contrary, such complications usually provide additional mechanisms for contamination to be detrimental to function and quality, the result usually being a requirement for more stringent control.

Wafer Fabrication

In this phase of the manufacturing, all the active and passive elements of the semi-conductor circuits are built onto or into the polished silicon wafer. On the microscopic level one can observe a cross-section of pure silicon being changed by: the addition of atoms deep into the pure metal to a controlled depth and concentration; metal being etched away; layers of silicon oxide being deposited over everything then selectively etched away to be replaced by silicon with other impurities, or aluminium; then layers being selectively etched away and replaced by more oxide and metal. The process is repeated until many layers are applied, interconnected, stabilized, passivated, etc.

Processes include: heat up to 1100°C; attack by exceptionally aggressive chemicals; flooding with chemical vapours so toxic that the most rigorous controls are required to protect people; exposure to violent levels of ionizing radiation and superheated, ionized plasmas. All of these take place to create circuits and active devices whose dimensions range down to 0.01 μm for critical features. The active elements which are interconnected into one device can be in the millions. All of these are packed into a chip measuring about one centimetre square. As all the dimensions are controlled by photographic processes (photo-lithography), the success or failure of manufacturing is absolutely determined by the control (and elimination) of contamination. The minimum class required is ISO 5 (Class 100), with the most critical steps requiring ISO 3 (Class 1) or better.

The processes encountered in wafer fabrication are given in Table 3.2.

The above steps are not necessarily in sequence, or all-inclusive, but serve to illustrate the complexity and enormous number of steps required. It is recognized that this sequence, or parts of it, may be repeated many times for one device. The complexity is added to by the precise inspections, measurement and cleaning required at least once between steps.

TABLE 3.2

Start with	Into	Produces
a. Polished wafer	Diffusion furnace or epitaxial reactor (high temperature atm. with water vapour)	Wafer with SiO_2 layer on surface
b. Oxide covered wafer	Spinner-coater which applies photo resist (PR)	Wafer with oxide PR layers
c. Wafer with PR	Pattern application (exposure of PR to UV light or electrons through photomask)	Exposure of pattern on PR
d. Exposed wafer	Developer (wash out either exposed or non-exposed pattern depending on type of PR)	Wafer with PR pattern
e. Wafer/PR pattern	Etch (etchant removes oxide exposed by pattern development)	Metal exposed through oxide
f. Wafer with pattern	Diffusion furnace or ion implant (add impurities to exposed silicon)	Implanted or diffused pockets in Si
g. Diffused wafer	PR removal (either plasma or wet etch removal)	Cleaned surface ready for more processing
h. Clean wafer	Reactor (deposit silicon layer)	New layer of active silicon
i. Metallized wafer	Repeat steps b to g	Wafer with clean surface
j. Clean wafer	Sputter reactor (apply aluminium coating)	Conducting layer
k. Metallized wafer	Repeat steps b to g	Clean surface
l. Cleaned metallized wafer	Repeat steps a to g	Clean surface
and so on until:		
x. Cleaned wafer	Reactor (nitride overcoat application)	Surface protected wafer
y. Surface protected wafer	Back etch (for thinning)	Thinned wafer
z. Thinned wafer	Backside plating	Electrical contacts

In the wafer fabrication area the objective of contamination control is to protect the work-in-process from errors created by contamination. There is a rule of thumb which says that the maximum size of particle which can be tolerated is one tenth the dimension of the smallest critical feature. For example, an electrical gate whose smallest critical dimension is 1.0 μm, can be made defective by a particle 0.1 μm and larger. Other types and sources of contamination, such as dissolved contaminants in deionized water, can produce killer defects that may or may not be traceable.

For the above reasons, in the wafer fabrication area the utmost in contamination control must be extended to the work until the wafer is completely protected.

Assembly and Test

This third phase is where the individual devices are tested, separated from the others on the wafer, mounted onto a substrate or leadframe, electrically connected from the terminating pads on the silicon chip to the leads on the leadframe, encapsulated, finally tested and shipped. In general, cleanliness levels between ISO 7 (Class 10 000) and ISO 5 (Class 100) are required.

The operation is as shown in Table 3.3.

The needs for contamination control in this phase are very different from the other two previously described manufacturing processes. Here the device is already protected from atomic and ionic pollutants, but must still be protected against large conductive particles shorting between leads. Also they must be protected against

TABLE 3.3

Start with	Into	Produces
a. Completed wafer	Multi-point prober (function test of each device)	Tested wafer. Failed devices identified with ink dot
b. Tested wafer	Mounting fixture (wafer mounted on plastic mounted on metal ring)	Wafer ready for dicing saw
c. Mounted wafer	Dicing saw (diamond saw precisely cuts through wafer to separate dies)	Separate die, adhesively held to plastic
d. Separated die	Die attach (vacuum probe picks off good dies and attaches them to leadframe)	Device to leadframe
e. Device on leadframe	Lead bonder (gold wire attachment of device to leads)	Device electrically attached to leads
f. Electrically completed device	Encapsulating (either plastic moulding or hermetically sealed 'can')	Completed device ready for final test
g. Completed device	Burn-in test (prolonged test under operating conditions and environmental extremes)	Device ready for shipment

electrostatic charge build-up and discharge (in this region the devices are very vulnerable, a 12 V discharge can destroy a circuit). Any oil or other material on surfaces may prevent sealing or adhesion of plastic or ink. Films or particles on a surface may also interfere with obtaining a good electrical contact and cause false readings on test. The principal difference is in the size or quantity of contaminants. In the wafer fabrication area contaminants that are too small to be seen in an inspection microscope may be fatal. The opposite is true in assembly and test; the fatal particles or other contaminants may be too large to be readily picked up in microscopic or other inspections.

Thus we have a brief description of the semiconductor manufacturing process and why contamination is a problem. It has been simplified to illustrate how important the problem is. We must now consider how to design clean areas suitable for the manufacture of semiconductors.

DESIGN GUIDELINES

Wafer fabrication facilities are among the most expensive to build and operate. To make matters even worse, the expensive state-of-the-art equipment can be superseded within two years and the average wafer fabrication area can have a maximum useful life of five years. These observations demonstrate that wafer fabrication must produce enough profits to amortize the cost of construction in the first two years, or the probability of success is small. It is important therefore to keep the goal of reducing these high construction and operating costs firmly in mind as one plans the design for the cleanroom.

The following guidelines are given to aid in reducing these costs:

Do not overdesign. A full vertical-flow, through-the-floor air return, ISO 3 (Class 1) cleanroom undoubtedly provides the cleanest, most versatile, easiest to use (most forgiving) facility. It can also be guaranteed to be prohibitively expensive to build and operate. If you can limit the strictest control to the area in which the work-in-process is most vulnerable to damage from contamination, and isolate that zone by a fixed barrier (such as a glass or plastic wall) then you may reduce the most costly area to 5% of the total.

Design for flexibility. Whatever the system may be, strive to make it possible to rearrange equipment, walls, filters, air returns, utilities, etc. easily. This should be at the least expense in the long term and, most importantly, cause the least disturbance to production when changes are made. You can be sure that equipment and processes will change and it is also axiomatic that the new equipment or process will not fit in the old space and will require a change in utilities, so there must be sufficient room for expansion.

Provice bulkhead-mounting of the process equipment, wherever possible, in critical contamination-sensitive areas. This allows separation of the product-flow from the process equipment and operation, maintenance and engineering personnel. The product movement area (highest level of cleanliness control) can hence be very small compared to the operations and equipment area (lower level of control). Major savings may be made here. Ideally, material handlers, properly trained and clothed, should be the only persons exposed to the same environment as the product. Then, should economic studies prove feasibility, it should be easy to replace those material handlers with an automated system. Currently, it makes little difference to contamination levels whether properly garbed and trained people or robotic systems are used and it is doubtful that this will change.

Provide a cleanroom environment only where it is needed, and only to the level needed. To do this successfully, it must be remembered that a 'clean zone' is like a vacuum—nature abhors it! The cleaner the zone, the more consequential are small leaks and pressure differentials. Therefore, a buffer zone of clean air of a lesser degree of control, surrounding the clean zone, is always appropriate, as is a slightly lower pressure. Even entry and exit from an uncontrolled area through an air lock creates an opportunity of contamination to enter.

Always examine process equipment. The process equipment should be cleaned thoroughly, and tested for residual contamination as well as contamination generation or retention, before introducing it into the cleanroom. Obtain a certificate from the manufacturer guaranteeing the maximum level of contamination dispersed from his equipment in terms of particles per wafer per pass, or other significant measurement appropriate to the equipment. This manufacturer's guarantee should be regarded as an extremely significant condition for purchase, as these contamination sources have enormous yield consequence.

Design utility distribution systems to:
1. Provide ready access for attachment equipment without the necessity for shutting down the system or cleanroom.
2. Provide adequate flow rates to prevent stagnation and impurity pick-up.
3. Prevent pressure fluctuation during operating 'runs'. Even slight variations can have disastrous effects on the product. It is true not only on the supply, but on the extract (exhaust) as well. It pertains to all systems of process utilities supply and return such as:

   - gases supply and extract (exhaust)
   - liquids supply and drain
   - electrical supply and earth-ground.

4. Periodically provide attachment points and blanked-off valves for all gases and liquids (except deionized water) so that a connection can be made without the danger of contaminating active process lines; they are virtually impossible to clean once contaminated. Deionized water systems should be provided with isolation valves and sterilization connections to:

   - make a new connection
   - perform sterilization
   - test to assure sterility and purity specifications
   - open the system.

5. Always design for the ultimate filtration of gases and liquids to be after the last valve or flow measurement device, so that the filter is the last thing before the product. This is so important that process equipment must be modified before installation if the situation is otherwise than required. There is no valve available at present whose operation does not produce particles in the fluid it is controlling, to a yield or function-damaging level.

Design for freedom from vibration and electromagnetic interference. These have become very critical. Both have some of the same origins and the design for the control of these should start, very early, with site selection. Before any site is adopted, thorough testing and analysis should be made of: seismic activity, soil conditions, proximity to railroads, highways, airport landing and take-off paths, adjacent industry, adjacent bodies of water where waves beating on shores can create earth impulses very difficult to attenuate and at a very damaging frequency. Large rotating electrical machinery in the vicinity produces both vibrational excitation of the ground and an electromagnetic field, the strength of which varies as the square of its power. High tension power lines in the vicinity not only produce electromagnetic fields but may excite some structural elements to mechanically vibrate at some multiple (or fraction) of the base frequency. All of these, and others, should be evaluated carefully by experts familiar with the microelectronics industry's needs before a decision is made to proceed. The time and money spent doing this may prevent the expenditure of correspondingly larger amounts of both, should a wrong choice be made. Certainly it is needful to establish parameters for the design of foundations, columns, spacing of structural elements, etc.

Define how the cleanroom is to be built. The success of a cleanroom project depends to a large extent on what is put on paper and how it is constructed. One feature, often ignored, but certainly within the purview of the designer, is the designation of when and how cleanliness controls are to be enforced during the construction phase. It is possibly the most important instruction the designer can give, for it determines in large measure the amount of built-in dirt that can greatly prejudice future operations. Dirt, built in during construction, will come into the process during operation and can determine success or failure of the cleanroom. My recommendation is to require cleanroom procedures to be applied as soon as the building is 'dried in', i.e. before any facility or utility systems or equipment are installed. From that point on, continuous cleaning is required, so that no dirt is trapped during construction to cause trouble later. Such cleaning should include: vacuum cleaning (with a brush) every surface, damp wash with free-rinsing, non-ionic detergent, rinsing with deionized water and a final wipe-down with 'tack-cloth' prior to turning the facility on. Following such a procedure should allow the specified conditions to be met within seconds of system turn-on and with few subsequent excursions.

DESIGN FEATURES

Layout

The design of semiconductor cleanrooms has evolved over several years. The design of a cleanroom which has been popular for a number of years is shown in Figure 3.1. The

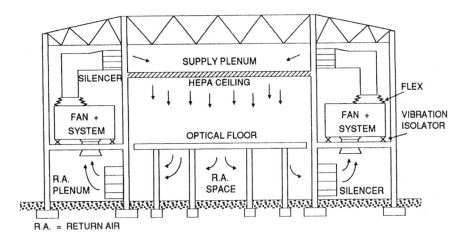

FIGURE 3.1. Vertical flow cleanroom often used in semiconductor manufacturing.

air flows in a unidirectional way from a complete ceiling of high efficiency filters down through the floor of the cleanroom. Some designs return the air through a plenum just below the floor, while other designs (similar to the type shown in Figure 3.1) have a large basement beneath the plenum, that basement being used for services. The design shown in Figure 3.1 is often called the 'ballroom' type because there is one large cleanroom. Typically it is over 1000 m^2 in floor area. It is expensive to run but it is very adaptable.

In the 'ballroom' type of cleanroom a ceiling of high efficiency filters provides clean air throughout the whole room irrespective of need. It is clear that the best quality air is necessary where the product is exposed to airborne contamination, but that lesser quality would be acceptable in other areas. Using this concept, less expensive clean-rooms have been designed in which service chases with lower environmental cleanliness standards are inter-dispersed with cleanroom tunnels (see Figure 3.2).

The production machinery is bulkhead-fitted so that the piped services can be run

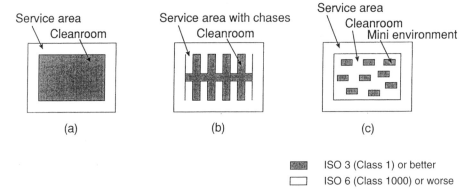

FIGURE 3.2. Plan views of three types of cleanrooms: a) ballroom type; b) service chase type; c) minienvironment type.

within service chases to the machinery. Service technicians can gain access to machinery without entering the clean space where the semiconductor wafers are exposed. It is also possible in the ballroom type of design to divide up the 'ballroom' with prefabricated walls and provide clean tunnel and service chases; these walls can be dismantled and reassembled in a different configuration should the need arise.

Service chases are normally supplied with a lesser quantity of air and hence have a lower standard of air cleanliness. The cleanroom tunnel can be designed to provide different air standards. Unidirectional air can be supplied from a 100% coverage of filters in the ceiling or from unidirectional flow work systems; in systems which do not require to be of the highest quality, air supply can be supplied by less than 100% filter ceiling coverage.

Various designs of this type have been used. Shown in Figure 3.3(a) and (b) are two typical designs. These are designs that have been used in the past but are still applicable in manufacturing areas or laboratories where less than state-of-the-art components are produced.

To achieve lower standards appropriate to less stringent contamination control areas in both sub-divided 'ballroom' and tunnel designs, the air supply volumes can be lowered by reducing the ceiling filter coverage from less than 100%. This method is shown diagrammatically in Figure 3.4.

If this method is to be employed, use may be made of Table 3.4. Table 3.4 is published in the Recommended Practice 012 of the IEST although the nomenclature of the room classification has been changed to that used in this book. Please note that the values used in this table are only a guide and are considered by some authorities to be inappropriate for cleanrooms used in manufacturing industries other than semiconductor.

If the cleanroom design uses an air supply plenum, then the unfiltered air in the plenum will be at a higher pressure than the air in the cleanroom. Unfiltered air can therefore leak from the supply plenum into the cleanroom through badly sealed, or unsealed, joints in the structure (Figure 3.5). Particular care must therefore be taken to ensure that the joints are correctly sealed. Such leak problems can be overcome if the area above the ceiling is below the pressure of the cleanroom. This can be achieved by

TABLE 3.4. Air velocities in cleanrooms.

Class	Airflow	Average velocity (ft/min)	Air changes per hour
ISO 8 (100 000)	N/M	1–8	5–48
ISO 7 (10 000)	N/M	10–15	60–90
ISO 6 (1000)	N/M	25–40	150–240
ISO 5 (100)	U/N/M	40–80	240–480
ISO 4 (10)	U	50–90	300–540
ISO 3 (1)	U	60–90	360–540
Better than ISO 3 (1)	U	60–100	360–600

$$\text{Air changes per hour} = \frac{\text{average air flow velocity (taken over the whole supply ceiling)} \times \text{room area} \times 60 \text{ min/hr.}}{\text{room volume}}$$

N = non-unidirectional; M = mixed flow room; U = unidirectional flow.

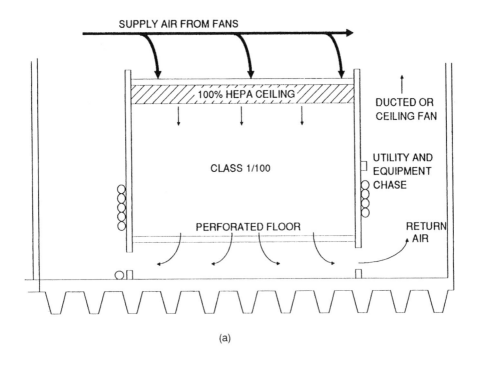

(a)

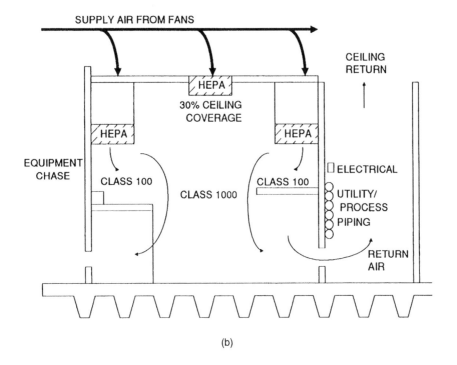

(b)

FIGURE 3.3. Two types of tunnel and service chase designs.

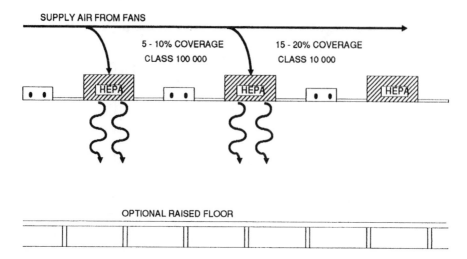

FIGURE 3.4. Reduced ceiling filter coverage.

individually supplying the filter housings with ducted air or by using individual fan/filter modules; these methods are often used when the filter coverage is less than 100%.

A reduction in capital and running costs of a semiconductor cleanroom is always sought, especially if this is accompanied by an increase in yield brought about by enhanced contamination control. There has therefore been much interest in what have been variously called 'isolators', 'barrier technology' and 'minienvironments. Minienvironments is the term commonly used in the semiconductor industry.

A minienvironment uses a physical barrier (usually a plastic film, plastic sheet or glass) to isolate the susceptible or critical part of the manufacturing process from the rest of the room. The critical manufacturing area is kept within the minienvironment

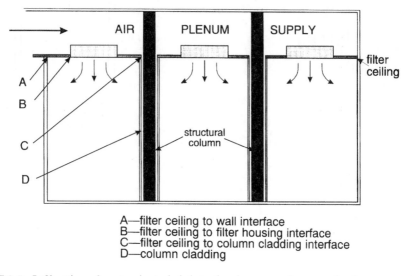

A—filter ceiling to wall interface
B—filter ceiling to filter housing interface
C—filter ceiling to column cladding interface
D—column cladding

FIGURE 3.5. Infiltration of contaminated air into the cleanroom from an air plenum supply.

and provided with large quantities of the very best quality air, the rest of the room being provided with lower quantities of air. Shown in Figure 3.6(a) is a diagram of the air supply design used with a more traditional ventilation method. In this design large quantities of a unidirectional flow of air are provided to those parts of the room where the production personnel move wafers from machine-to-machine and lesser quantities of air are provided for the chases where the bulkhead-fitted machines are serviced. Shown in Figure 3.6(b) is a diagram of the air supply design used with mini environments, where the highest quality of environment is provided within the minienvironment and a lesser quality of environment is provided in the area where the production personnel move about as well as within the service chases. The total air supply volume can be seen to be much less when minienvironments are used. It can also be appreciated that this type of system lends itself to automation and the use of robots: that may or may not be desirable.

As well as using minienvironment to isolate the area where the silicon wafers are exposed, the wafers can also be transported between processing machines in specially designed carriers which interface with machines through a Standard Mechanical Inter-Face (SMIF). The wafers are then laded by a SMIF arm into the processing machine where it is contained within a minienvironment. After processing, the wafers are loaded back into the carrier and taken to the next machine.

Shown in Figure 3.7 is a plan of a semiconductor production area which would be used for VLSI-CMOS or more critical product manufacturing and is designed for use with minienvironments. Typical requirements for the air cleanliness within the area would be similar to the following:

1. An environment where the wafers are fully exposed—ISO 3 (Class 1) or better.
2. An area where wafers are protected in cassettes and enclosed boxes—about ISO 6 (Class 1000).
3. Machine–technicians–engineering chase area—ISO 6 to ISO 7 (Class 1000 to Class 10 000).
4. Adjacent areas outside the cleanroom envelope—standard air conditioning.

In Figure 3.7 the areas where the wafers are exposed are isolated by rigid walls of plastic or glass and provided with a unidirectional air supply which will give ISO 3 (Class 1), or better, conditions. These minienvironments are shown in the darkest shade. The darker grey area represents the area where the production personnel move the wafers which are protected in cassettes and enclosed boxes; this may be ISO 6 (Class 1000). The lighter grey area is the service area where technicians gain access to the services and machines; this may be ISO 6 to 7 (Class 1000 to 10 000).

The particular layout shown in Figure 3.7 was designed for use with the Standard-Mechanical-Interface-Format (SMIF) system (Workman and Kaven, 1987). This concept, supported by the Semiconductor Equipment and Materials Institute, provides a standard entrance portal, thereby allowing a single type of handling equipment to introduce and remove a set of silicon wafers into the process equipment. When the process is complete, people or robots can remove the sealed SMIF pods containing the wafer cassettes from one machine and transfer it to another machine to be loaded by the SMIF arms (see Figure 3.6 (b)). The SMIF equipment is designed to optimize production yields and permit high-volume throughput, but the same layout concept

(a)

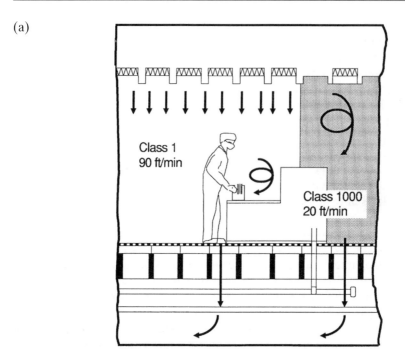

(b)

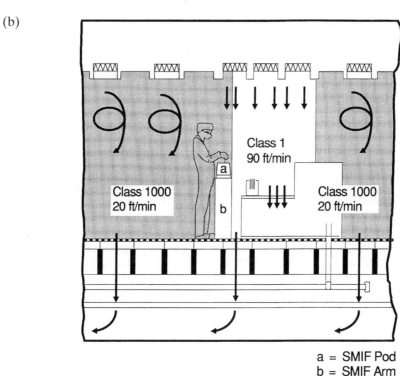

a = SMIF Pod
b = SMIF Arm

FIGURE 3.6. A comparison of the SMIF isolation and the more traditional approach. (a) Traditional type of ventilation. (b) Minienvironment using SMIF technology.

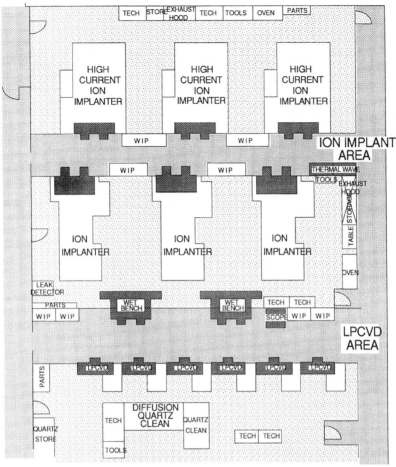

TECH = Technician's bench
LPCVD = Low pressure chemical
          vapour deposition
W I P = Work-in-progress

FIGURE 3.7. Wafer fabrication area designed for a SMIF/minienvironment system.

shown in Figure 3.6 (b) can be used with other systems. Other minienvironments, with different methods of accessing the wafers into the production machines, have been developed and as long as these are well designed, particularly with respect to the container for the wafer cassettes and the port which allows the transfer of the wafers into the minienvironment, they will work well.

Airborne conditions and surface particle contamination of the wafers produced in minienvironments are normally better than those achieved in the ballroom or tunnel/service chase cleanrooms. Wafer yields have been reported to have improved; the greater the density and size, the greater the benefit. A reduction in the costs of building a semiconductor fabrication area using minienvironments has been quoted to be less than 10%. However, in terms of the cost of such facilities, this can still be a significant amount of money. More important is the considerable savings to be made on running costs, as the air supply volume is likely to be reduced by about 50%. The cost savings will depend on the conditions thought necessary for the operator movement and service areas. ISO 6 (Class 1000) conditions have been suggested for he operator movement areas, but it has also been demonstrated that ISO 7 (Class 10 000) conditions do not influence the number of contaminants on the wafer surface. A downgrading of other requirements associated with lower cleanliness conditions in the operator movement area, e.g. clothing, gloves, wipes, cleaning and certification, will give a significant cost saving.

As well as the advantages quoted in the previous paragraph, there is also the potential advantage of a more economic start-up of a new cleanroom. Instead of opening up the whole cleanroom at once, it is possible to phase in, as required, the introduction of equipment. The contamination of the semiconductors during equipment installation is likely to be small in the minienvironment compared with the 'ballroom' type of area, where a phased plan would be more difficult. Savings in costs of using such a 'ramp up' of the production are likely to be large.

The factors influencing the design of microelectronic cleanrooms are often in conflict with each other and the resultant facility is always a compromise between cost, perceived need, convenience, individual preferences, highest level of technology and available funding. It is almost never of a single style or type, but of a multiple design configuration; this being an attempt to supply all of the needs of the process with one facility.

Air Flow—Direction

In most cases, for the maximum control of product cleanliness, the air flow should be vertically downward and should meet the ISO 3 (Class 1) or better cleanliness specification as it flows onto the product. Obviously, vertical flow onto the product is not possible in an evaporator, reactor or diffusion furnace. In those special cases, supplemental filter-blower units, scavenging air vanes, etc., should be tried. In any event, such specialized treatments should be qualified by thorough testing before adoption.

Some special environmental chambers which use a horizontal airflow or an airflow 45 degrees to the horizontal, have been quite successful in reducing the number of particles reaching the product in some production equipment. Such equipment, if available, may be worth investigating.

ULPA filters 99.9995% efficient when tested at manufacture against 0.1 micrometre-sized particles, and tested and certified at installation with a suitable particle challenge, must be used for ISO 3 (Class 1), or better. To permit ultimate flexibility and minimal record keeping of what kind of filter goes where, and because the differential cost between 99.00% and 99.9995% efficient filters is small for the convenience given, it is recommended that the same higher grade type of filters be used throughout the cleanroom. These can be used without any protective or ornamental screen or 'eggcrate' on the down-stream side of the filter. Internal or inlet dampers should be used to regulate air quantity.

The air should exit the room through perforated panel floors with integral air dampers to regulate the air flow. The floor covering should be made from a conductive, high pressure laminate and the system must be grounded (see Chapter 9). ISO Classes 3–5 (Classes 1–100) areas will require 100% coverage by perforated panels, while ISO

Classes 6 and 7 (Classes 1000 and 10000) require only 50% or less coverage by perforated panels. Some solid panels will be required for use under equipment which has a closed base (it is easier to cut utility piping holes in solid panels).

An alternative, acceptable solution to the design of air flow for ISO 6 and 7 (Classes 1000 and 10000) is to use classical turbulent ventilation, the return air vents being in walls or islands and close to floor level. This may or may not pose a problem for ducting the return air back to the recirculating air systems, depending on over-all design. With this system one does, perhaps, give up some of the inherent flexibility of the total vertical flow, through-the-floor return system. Upgrading of such random-flow (turbulent) systems to ISO Classes 3–5 (Classes 1–100) or above would involve a prolonged shut-down and complete revision of the entire area; with the vertical unidirectional flow it is more easily and quickly accomplished.

Air Flow—Quantity

ISO Classes 3–5 (Classes 1–100) should be designed to between 0.3 and 0.5 m/s depending on the flow through the respective clean zones. It is important to observe the effect of air flow through critical production-equipment interface-locations. There are occasions when a reduction of airflow can result in a significant lowering of the particle level or reduction in the machine interference (cooling). Classes ISO 6, ISO 7 and ISO 8 (Classes 1000, 10000 and 100000) should be designed for 50, 30 and 18 ft/min (0.25, 0.15 and 0.09 m/s) flow through them. Primary control is achieved by providing equivalent filter/solid panels in the floor. After installation and start-up, adjustment of filter and perforated-panel dampers will be used to fine-tune air flow direction and quantity.

Airborne Molecular Contamination

It has been demonstrated that airborne molecular contamination has an effect on semiconductor yield. As the size of device geometry reduces, this is likely to be viewed with increasing concern. At the time of writing this chapter, the importance of the various types of molecular contamination, with respect to the various production steps, has not been fully determined and hence the requirements for the different areas in the semiconductor facility is not clear. Another problem that exists is that analytical methods for measuring airborne molecular contamination are insufficiently accurate, or available, to give users or designers a clear insight into the problems and solutions.

A standard for molecular contamination has been produced by SEMI and is known as SEMI Standard F21-95. This considers molecular contamination in four groups. These are A(cids), B(ases), C(ondensables) and D(opants). The classification is given an 'M' nomenclature followed by the type of molecular contamination being considered (A, B, C or D) and then the concentration in parts per trillion (molar). Thus, a classification of MA-100 would set a limit of 100 ppt of gaseous acid.

The most common airborne molecular contamination in the cleanroom is hydrocarbon in nature but can also include acids, bases and other process chemicals. The molecular contamination found within a cleanroom will come from:

- The fabric used in the construction of the cleanroom;
- The machinery within the room;

- Uncontrolled chemical releases;
- The people within the room;
- The outside make-up air;
- The air conditioning system.

The architectural and structural components of the cleanroom can be a major contributor to airborne molecular contamination. Many of the conventionally-used components and materials will be a source, including items such as gels, caulks, sealing compounds and common protective coatings such as primers and paints. Some plastic pipes and ducts can outgas and will therefore be a source. Flooring can be of special concern. The area of floor will be large and, depending on the materials used, could outgas considerable amounts of airborne molecular contamination. It will therefore be prudent to minimise molecular diffusion from the building fabric by use of materials that have a minimum of out-gassing. Air extract systems should be designed to effectively control the dispersion of contaminants from machinery, as well as minimise the dispersion of process chemicals both under normal working conditions and where an accidental spillage may occur.

As molecular contamination is in a gas phase, it will be recirculated round the air conditioning system. Air filters packed with activated carbon, or activated alumina, will reduce the levels of contamination. Activated filters are not very efficient, being about 99%. This efficiency is much less than that available from airborne particles filters but appears to be adequate for the present requirements. As knowledge is gained on this topic and more effective methods are devised to measure molecular contamination, then it is likely that more effective removal methods will be available in the future.

At the time of writing, the control of airborne molecular contamination is not an established measure for all new semiconductor facilities. However, where it has been decided that control of molecular contamination is not necessary, it would be prudent to include the flexibility and space to include control measures in the future. If mini-environments are used in the facility design, then control of molecular contaminants can be built into their air handling units. This is a more economic solution than controlling the recirculated and make-up air of the whole facility.

Filter suspension system

Regardless of whether a pressurized plenum system, or a ducted filter system is used, the incorporation of filters, solid panels and lights into a stable, non-leaking, unified system is dependent on the integrity of the sealing system. The one which has been found to be the most reliable has been the fluid- or gel-seal system of which there are several varieties. Properly installed and tested, they should not be a source of trouble. The suspension system should contain the fluid (or gel) while the filter has the knife-edge which mates into the fluid (or gel). (See Figure 8.18 in Chapter 8.)

Recirculation Air Moving System

Shown in Figure 3.8 is a drawing of a typical air movement system used to recirculate the air passing through the floor grilles to the high efficiency air filters. This system should have the following characteristics and requirements:

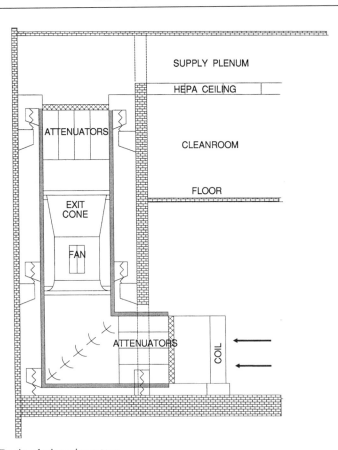

FIGURE 3.8. Recirculation air system.

1. Fan type: This should be an axial vane type with an optimum transition type of attenuator on both the inlet cone and exit. The motor should be external to the air stream.
2. Structural support for fans should be out-board of the wafer fabrication area with separate foundations, detuned column support and maximum isolator design, so that transmitted (structural) vibrations are kept to a minimum and not within the 0–200 Hz range.
3. The fan chamber should be lined with perforated stainless steel (type 304) sound attenuators stuffed with high density polypropylene-bagged fibreglass batting. This attentuation should be carried past the fan chamber so that at least one 90 degree bend of the air path on the upstream and downstream side is so treated.
4. The fan–coil–ductwork–perforated-floor–filter–prefilter system should be designed to produce a system pressure drop of not more than 2.0 inches pressure (500 Pa) with clean filters fitted.
5. The fans and motors should be remote from the wafer fabrication area so that EMI from the motors does not interfere with the sensitive fabrication process and inspection equipment. The allowable proximity varies directly as the square of the horsepower.

6. Because all of the air passages are so large, they require turning vanes and, depending on design, sound attenuating treatment. If a pressurized plenum system is used, the entrance into the plenum requires baffling to prevent backflow through the filters, caused by air-shear across the top of the filters. The depth of the plenum should be such as to permit even distribution of the air to the upstream side of the filters, with the air flow across the filters being in the nature of a pressure front rather than a velocity front. In general, the goal is to hold air velocity horizontally above the filters to less than 1000 ft/min (5 m/s). As an example, for a room 100 ft (30 m) wide, with air fed from both sides, the plenum should be a minimum of 5 ft (1.5 m) deep. Velocity of the air, then, would be a maximum of 1000 ft/min (5 m/s) at the sides, tapering to 100 ft/min (0.5 m/s), in the centre.

7. Cooling coils should be mounted upstream of the fans. Direct digital control is preferred with the temperature sensors (electronic) located after the fans. The cooling system should be a dual-function type, with the cooling coils controlling only sensible heat.

Fresh Air System

Fresh air for the wafer fabrication area should be taken directly from the outside, and conditioned for use. The system must have sufficient capacity to purify (filters, absorbers, washers) and control humidity to provide the basic latent heat control for the entire fabrication area. This system should also have direct digital control, which may be by either a face-and-bypass system, treating only a fraction of the air supplied by the fresh air system, or a full-flow system. In any event, controls measuring the relative humidity of the cleanroom should be used by the microcomputer to call for moisture (usually electrically generated steam) or lower temperature (refrigerated or absorption drying) to adjust the total latent heat. The fresh air requires to be distributed evenly to the intake of all the recirculating air fans. One of the easiest means to accomplish this is to run a distribution duct the full length under the floor so that the fresh air can flow evenly toward the recirculating fans mixing with the recirculating air as it goes. This is possible, of course, only if enough space exists under the floor to do this.

The control system which provides the finest control is that which uses heat to make the final adjustment of both temperature and relative humidity rather than cooling because the incremental change is easier to control.

For ISO 5 (Class 100) areas or better, it is imperative that the final filter should be at least an HEPA filter (99.995% efficiency). It has been shown that if this is not done, the majority of particles smaller than 0.5 μm diameter in the cleanroom, come directly from the outside through the fresh air system.

Air Return

The most versatile and easily controlled air return system is through-the-floor. If enough space can be provided under the raised floor, it is ideal to install all of the utility distribution there, bringing it to the process equipment from below.

The advantages to be gained from through-the-floor air return are:

1. It is difficult to block the air return. With side-wall return the production equipment and furniture almost always severely restrict the air flow.
2. It is easy to segregate and totally isolate the critical operations, but still have them accessible.
3. The most vulnerable operations, where silicon wafers are fully exposed, are best behind glass or plastic and totally protected by a well controlled flow of unidirectional air from the ceiling and down through the floor.
4. The simplest automation system presently available (SMIF) is readily adaptable and available on a variety of production equipment and offers total protection of wafers in ISO Class 3 (Class 1) conditions when outside the production chambers.
5. With the newest, most cost-effective layouts, the side-wall return is very difficult to use. It would require either a double-wall air return plenum between the ISO Class 3 (Class 1) zone and the ISO 6 or ISO 7 (Classes 1000/10 000) zones, or the use of the entire ISO 6 or ISO 7 (Classes 1000/10 000) zones as an air return plenum. The second of these alternatives is not attractive because of the non-controllable air flow in that zone. The same problems are also inherent with utility supply and return. All systems that one can conceive are awkward in some measure if the supply and return of gases, liquids and electrical systems have not the under-floor space for distribution.
6. Through-the-floor air return can be achieved in one of two ways:

    (a) The under-floor can be designed as a basement giving full structural support to the perforated floor.
    While it is expensive, it is not generally as costly as adding space laterally. In addition, many of the utility supply and disposal systems may be housed in this space. Should this alternative be adopted, consideration should be given to providing a good clearance—say 18 ft (5.5 m) between basement floor and the fabrication floor. In this amount of space there is room to do all of the desirable things such as relegation of most of the process equipment maintenance and supply functions to that area. Also, most of the maintenance work involved in adding or moving process equipment may be carried out without affecting production,
    (b) Install a raised access floor on top of the building floor.
    This is a satisfactory solution provided enough thought is given to the additional usages planned for the under-floor space. It should be deep enough to accommodate the air flow without significant increase in air resistance (it should be no greater than the pressure drop through the floor), while at the same time allowing placement of necessary process equipment support items such as filters, regulators, transformers, and the required utility supply and distribution system.
    The raised floor system needs to be of a heavy-duty design (350 lb/ft^2, i.e. 16.8 kN/m^2 minimum) with

    - full stringer support
    - four bolts to the pedestal
    - locking levelling nuts

- alternate row Z-bar bracing both ways
- either epoxy-cemented or bolted pedestal attachment to the floor.

The concrete floor under the raised floor should be coated, after pedestals are installed, to provide acid-resistant, water-proof protection to the concrete. In addition, sensing systems must be installed to detect and provide an alarm when any liquid leaks occur.

The total raised floor system must not create any amplification of the basic floor and structure vibration. Indeed, installed correctly, it should dampen the basic vibration significantly, since the structure being tied to the floor converts it from a vibrating membrane to a box-beam.

Fire Protection

Because of the use of solvents, gases, and other flammable materials which are often used at the high temperatures and pressures inherent in wafer fabrication facilities, the usual fire risk category is 'extra-hazard'. This requires a maximum-control sprinkler system (minimum spacing), both above and below filters, and also below the floor if flammable gases or liquids are piped through that zone. The alternative is a carbon dioxide or 'Halon' system, generally considered too expensive for large non-confined areas. These are usually reserved for under-the-floor high risk areas.

Sensing of fires requires line-of-sight coverage of the entire wafer fabrication area with both infra-red and ultra-violet detectors, and the under-floor area as well. The control system must be set to permit detection of, and alarm for, both smoke and fire with a time delay to make certain the emergency is not a false alarm, prior to triggering action, The delay permits evacuation of personnel from the affected zone before the suppressive action is taken. Also of significant importance is the fact that any interruption in production is extremely costly. A deluge of water on operating equipment, its microprocessor controls and electronic sensors; the shock of super-cold carbon dioxide impinging on heated process equipment; even the instantaneous shut-down of high thermal-inertia equipment like diffusion furnaces, CVD reactors, etc. can cause enormous damage.

Walls

Those used in a wafer fabrication are basically divided into two types:

1. *External walls*. These range from vault-type construction to standard architectural structures. Either may be appropriate depending on the type of product to be manufactured. Included in this category are those which are used for security purposes and provide the ultimate in radio frequency shielding (Figure 3.9). They provide up to 100 dB attenuation of internally transmitted signals or inadvertent transmissions over telephone lines, power lines, plumbing systems, extract systems, and radios, etc. Also to be considered is protection of computer controlled equipment against electromagnetic noise (Figure 3.10). The internal finish of peripheral walls should be a baked enamel or porcelain surface on metal. Metal panels must be sealed at all joints.
2. *Internal walls*. In the wafer fabrication area it has been found desirable, wherever

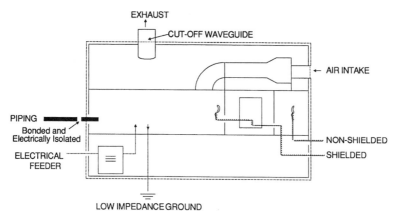

FIGURE 3.9. RF shielding.

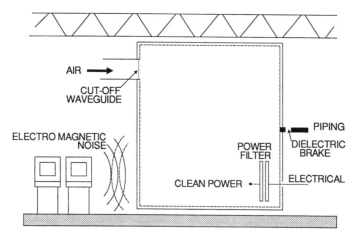

FIGURE 3.10. EMF shielding.

possible, to provide 'store front' walls composed of anodized aluminium structural members with glass or polycarbonate panels (which have been given a hardening treatment to reduce surface abrasion). The glass or polycarbonate panels may be replaced by metal panels where equipment penetrations are necessary. This type of structure permits see-through to all operating and maintenance areas and stimulates the required continual cleaning necessary for proper operation.

Lighting

In the wafer fabrication area lighting is best handled by fluorescent tubes mounted on the filter face as an integral part of the filter. Ballasts must be mounted remotely, out of the cleanroom. To preclude problems of deciding where the yellow and white lights should be, and to allow for future revision of layout, it is best to make all of the lights yellow. A guarantee should be required from the supplier stating that the yellow tubes meet the requirements for semiconductor manufacturing.

Electrical System

The electrical supply system is the utility system which is probably under the least control yet is the most important. Most of the process equipment is now micro-computer controlled. An interruption of a few cycles causes unprotected computer equipment to drop out, and production to stop. The question of how to protect against such events is a difficult one and most of the solutions are extremely expensive. Fortunately, most of the microcomputer-controlled equipment also has an uninter-ruptible power supply (UPS) which supports the control system for at least a few minutes. In a significant number of events these interruptions are of short enough duration to prevent serious damage, even to the product-in-process.

The most commonly used preventive measure is to install a loop supply, with feeders coming into the loop from entirely separate sources; even separate power plants if possible. The effectiveness of this is problematic, since the 'Grid-System' ties all of the sources together. Some have felt so strongly about the problem of power interruption that huge UPS systems have been installed. With this system the source power is fed into battery clusters which in turn feed generators or alternators into the internal loop, with the battery clusters having sufficient capacity for soft shutdown.

Regardless of the source power system, there are requirements, by law, for back-up power to important systems such as: toxic extract (exhaust) scrubbers, sensors for toxic or hazardous materials, alarm systems, and security systems. These are usually accom-modated by an emergency generator fuelled by diesel or natural gas. It is most impor-tant that they be exercised frequently to a strict schedule to assure readiness for emergency use.

Monitoring Control and Alarm System

The monitoring control and alarm system has a high priority in operating a wafer fab so it has great significance in design as well. The designer must consider:

- purity and physical condition of liquids, process gases and environmental air
- safety of personnel, plant and process equipment
- product yields and quality
- community relations with regard to effluent, toxicity or odours
- governmental agencies' requirements for control, measurement and reporting
- facility and utility system's condition and operating characteristics security.

It is obvious that no one system will do all of those things at once. The design of that system could be as complex as the devices being manufactured, or more so,

To make sure the system does not become excessively complicated, the following steps should be considered:

- Develop systems as simple and direct as possible to take care of one type of func-tion (preferably microprocessor controlled).
- Use the same system architecture and components (such as sensors, micro-processors, output signals, etc.) so far as is possible for each system.

- Use a simple multiplexer function to tie them all together in a central computer system which will deal with alarm, display, recording, trend analysis, preventive maintenance scheduling, mean time before failure analysis, forecasting, etc.
- Do not provide for resolution and analysis of data outside the computer system. It will never be done. Likewise, provide for printing of the data only when it is necessary for mailing or similar purposes.

The following alarm and control systems should be considered as the minimum required:

1. Toxic gas and liquid sensing in room air. The sensors should have a sensitivity an order of magnitude lower than the threshold limit value (TLV). TLV is that concentration judged safe for normal persons exposed for 8 hours. The sensors should process the signal real-time. No sensitized tape systems should be used. An alarm signal should:

   - Provide unmistakable, loud audible alarm in the facility as well as security headquarters
   - Provide visual alarm (flashing light)
   - Notify safety; security; facilities, fire department
   - Record all particulars on the facility's computer data bank
   - Shut off automatically the toxic gases and liquids at their storage area
   - Perform other automatic corrective functions such as turning on water sprays on leaking pyrophoric gases.

2. Cleanroom air monitoring for temperature, relative humidity, acid fumes, and solvent fumes is required in relatively few places in the cleanroom. Particle monitoring should be done very frequently in the air stream approaching the product, as close as it is practical. The average wafer fabrication area should have several hundred fixed sensing points to cover the product at its most critical stages.
   Very careful advance planning is needed to decide

   - what and when data are to be taken,
   - where and how they should be stored and for how long, and the reporting format and frequency.

   The computer should be programmed to provide a view of current conditions, a synopsis of recent data (especially excursions or failures) and a limit placed on how long data are to be stored before automatic erasure.

3. Gas impurity analyses from the continuous gas monitor system. These should analyse plant gases (including clean dry air), stored liquid gas sources, or cylinder gases. All may be fed into the computer and analysed by the same method as air monitoring data, i.e. immediate notification of excursions out of specifications or failures: on call current conditions; synopses of historical data; cut-off point for erasure.

4. Automated measurements of toxic or other specified harmful content of extracts (exhausts) and liquid wastes being discharged from the plant may be required by law. Such data collection, analysis, reporting, storage, etc., are usually specified

by the appropriate authorities. These may be fed to be processed, stored, and reported by the computer. The difference from the previously described systems is that legal requirements will dictate the length of retention of records and whether hard copy is required.

5. Deionized water has higher standards than it had very few years ago. Where once 18 megohms was considered the ultimate standard of water purity, now parts per trillion of iron, silica, hydrocarbons and chlorine are known to be yield-killers and their effect on resistivity of water is minuscule. Bacteria can have a profound effect on the quality of epitaxial layers, even their remains can reduce yields. These and other problems are discussed in Chapter 11.

   Tighter specifications for deionized water are required and more stringent manufacturing controls are required. Unfortunately, to date, some of these requirements cannot be assessed by in-line automatic equipment but must be analysed in the laboratory. As engineering advances, equipment will become available.

6. Fire detection system. This will be as detailed previously.

7. Security system. This system will depend on the type of product. In all cases, however, it is essential to couple the major monitoring, control and alarm systems into security headquarters. The purpose is to permit them to make an informed response to any emergency and communication with responsible parties.

Remainder of System

The remaining systems required for a wafer fabrication area deserve much more space than will be given here. Two out of three of the systems are treated in depth elsewhere in the book, i.e.

1. Deionized water generation and distribution, Chapter 11.

2. Gases, storage and distribution, Chapter 12.

3. Liquid chemicals supply, storage and distribution.

In the automated or semi-automated factory of modern design, all possible fluids are piped to the point of use and the spent fluid piped off for recovery or disposal. Since most of these fluids are hazardous to people, and most are as pure as it is possible to produce and determine by test, it makes good sense to do all of the handling, transporting, use and disposal in a closed system.

There are numerous firms, including the principal suppliers of the chemicals, whose speciality is the design, supply, installation and operation of such systems. Because the systems, chemicals and piping materials are in a constant state of development it is best to let those closest to the problem and solutions do the design for you.

4. Extract (exhaust) systems.
This is a common mechanical engineering design problem with uncommon requirements.

Generally speaking, the solvent and heat extract (exhaust) systems are straightforward, as long as the toxic gas systems are kept separate, and the solvent levels remain under control. Some of the materials used in the processes are carcinogenic and must be collected in the scrubber effluent and disposed of properly.

The acid extract (exhaust) system is also comparatively easy to design except for NOx which is very difficult to remove from the gas stream and HF where the scrubber liquid effluent must be collected and disposed of as a toxic waste material. Great care must be taken to prevent scrubber liquid effluent from the toxic gas systems mixing with the acid gas systems or the toxic gas will be regenerated. Arsine, diborane, stibine, phosphene, etc. can never be safely ignored. Toxic extracts (exhausts) are always handled one system to one system; they are therefore small. The extract (exhaust) should be passed through a high temperature oxidizing unit, or a strongly oxidizing liquid reactor. It should then pass into a high efficiency water scrubber and then to atmosphere through an externally mounted centrifugal fan. The liquid effluent of the reactor and the scrubber must be disposed of as a toxic material.

Some encouraging results have been obtained recently by use of pelletized-bed absorber-reactors through which the toxic extract (exhaust) is passed. The effluent is composed of carbon dioxide and water vapour. When instrumentation is reliable enough to provide confidence in prediction of break-through (escape of toxic gas), such devices will be a welcome addition.

5. Cooling water systems.
There are enormous quantities of cooling water required for a wafer fab.
    At least four types are required:

- Chilled water—normal, treated water (no chromates).
- Tower water 85°F (29°C) maximum) used in large quantities for process equipment cooling.
- 58°F (14°C) chilled water for cleanroom temperature control coils (dry).
- High resistivity cooling water for process equipment, especially that heated by RF induction. The temperatures and pressures vary from equipment to equipment so the size-requirement of these systems tends to be rather small.

Some suggestions may aid the design of cooling water systems:

1. Some production equipment cannot stand high internal cooling water pressure. Should the system differential pressure exceed the limiting pressure, install a properly sized pump and control valve between the equipment water outlet and the return line of the cooling system. This will permit the desired internal pressure, while allowing the differential pressure to do its work.
2. The 58°F (14°C) water for the dry coil cooling may be produced by using a recirculating (through the coils) system. This will require just enough of the 58°F (14°C) water continuously bled back to the return of the chilled water system to permit (40°F (4°C) chilled water to enter the loop to keep the temperatures constant. The load will vary slowly so that it is easy to hold the loop constant within 1°F (0.5°C). To install a water-to-water heat exchanger to handle this capacity would be difficult because of the large sizes required.
3. In northern climates it is often possible to use ground or surface water for cooling purposes. One factory in Germany was able to avoid spending over one million US dollars for chillers and saved a half million dollars a year in electrical power

reduction by using available ground water to replace all chilled water (except that used for dehumidification). It is important to note that risk of contamination of the source water is eliminated by installing a heat exchanger between it and the water used for cooling.

6. The material used for piping systems in wafer fabrication areas is discussed in Chapter 13 of this book.

CONCLUDING REMARKS

This chapter has presented the latest in design concepts for a wafer fabrication cleanroom and other systems required to make a successful facility. Other concepts have been alluded to, but, for lack of space, not detailed. There are many concepts and many variations of each. Clear thinking and sound judgement are the only way to choose between them.

It is quite true that enough is now known about control of contamination in environments that any conceivable requirement for such can be met, providing that instrumentation and methodology for test and measurement exist.

ACKNOWLEDGEMENT

Figure 3.6 is reproduced by permission of Asyst Technologies.

REFERENCE

Workman, W. and Kavan, L. (1987). 'VTC's submicron CMOS factory', *Microcontamination*, **5**(10), 23–26.

4 The Design of Cleanrooms for the Pharmaceutical Industry

G. J. FARQUHARSON and W. WHYTE

INTRODUCTION

This chapter considers the design, construction and commissioning of a pharmaceutical clean area. Chapter 2 of this book provides a bibliography of available standards and guides and other chapters provide the reader with information concerning key components that will be embodied within the overall pharmaceutical clean area. It is therefore proposed to use the information contained in these sections, and to develop other features and performance requirements where they are particularly relevant and important to the creation and operation of a successful pharmaceutical clean area.

Within the pharmaceutical industry, and, in particular, that part related to the manufacture of sterile pharmaceutical products, a single cleanroom is rarely encountered. Rather there are suites of rooms, integrating various grades or classes of rooms to accommodate different parts of the manufacturing process. It is also possible, particularly in the production sterile pharmaceuticals by an aseptic process, that isolators will be used as an alternative or complimentary solution to the clean area design.

The relationship of the pharmaceutical clean area within a total Quality Assurance Programme is important to understand. The design and function of the pharmaceutical manufacturing area forms a significant part of Good Pharmaceutical Manufacturing Practice (GPMP), these being the requirements laid down by Governmental Agencies and contained in various Guides to GMP (see Chapter 5). Provision of clean areas cannot provide all the appropriate conditions required, as other aspects of quality assurance associated with work practices are also very important. However, the clean area element is an essential and difficult part of the compliance with the Guides to GMP. Also of importance is the fact that the facility represents a significant capital investment.

The problems associated with designing clean areas for pharmaceutical manufacturing lie, to a large extent, in the diversity of requirements of the pharmaceutical manufacturer. If their requirements are not understood then it will not be possible to provide suitable clean areas. It is therefore necessary to consider what is produced in the clean areas and how this affects the design in terms of room layout and environmental quality.

Cleanroom Design. Edited by W. Whyte © 1999 John Wiley & Sons Ltd

TYPES OF PHARMACEUTICAL PROCESSES

There are many different types of products produced in pharmaceutical manufacturing areas. For the purposes of this chapter they may be first divided into:

- Injectables—those products that are injected into a person.
- Topicals—those that are applied to the outside of the human body.
- Orals—those that are ingested.

Generally speaking, injectables require much higher quality environmental conditions for production than topicals or orals, as the human body can cope with less bacterial contamination of the bloodstream than on the skin and in the stomach. Eye drops are an exception to this general rule as the eye is generally more susceptible to infection than the skin.

Injectables

In the context of clean area layout and environmental requirements, sterile injectable products can be divided into three classes, i.e.

- Aqueous products
- Freeze-dried products
- Powder products

Aqueous Preparation   The preparation of injectables in water (or as an emulsion) falls into two major areas, i.e. aseptic production and the production of terminally sterilized products. If the product is heat-stable then it will normally be terminally sterilized. These products will therefore have little bacteriological risk to the patient. However, if the injectable material cannot be terminally sterilized because it is unstable, as, for example, with human hormones and vaccines, then the product must be assembled aseptically. This means that the liquid is filtered free of microorganisms and filled into a sterile container in a clean area which is as free of microorganisms as is reasonably possible. Aseptic production is therefore the most exacting process in pharmaceutical manufacturing and difficult to do well. Class 100 conditions are required in the critical area where the product is exposed and this would usually be achieved by provision of a unidirectional flow system or isolator within a conventionally ventilated room.

Freeze-Drying   If a pharmaceutical product is unstable in solution then it may be freeze-dried (lyophilized). The product is made up in a solution which is filtered free of bacteria and dispensed into sterile containers (usually vials or ampules). The containers are then taken to the freeze-drier and the water removed by the lyophilization process. They are then taken from the freeze-drier and capped or closed. They are not terminally sterilized but microbial growth will not occur in the dry powder. Freeze-dried products therefore come between aseptically and terminally sterilized pharmaceutical products in the categories of risk. However, in the view of many Regulatory Authorities they are an aseptic product and must be produced in conditions with the same standard as required in aseptic filling rooms for the production of aqueous products.

Powder Filling   In this process the product is delivered as a powder to the filling room and dispensed, by machine, into the containers and the containers closed. Antibiotics are very often filled in this way. No terminal sterilization is carried out.

Powder will not support the growth of bacteria and if the powder is an antibiotic the chance is even further diminished. However the Regulatory Authorities are likely to regard them as aseptic products.

The production of powder-filled products has two special problems. The humidity of the air must be kept low to ensure the flow of the powder and powder filling may give rise to clouds of particles which, if the product has high potency, will give problems of personnel contamination. Special consideration must therefore be given to extract systems for the filling machinery, containment cabinets or isolators and special air movement control schemes. It is considered by many authorities that a separate building is required for the production of these biologically active powders, or at least a dedicated facility which is effectively isolated from other production areas.

Topicals

These products are applied to the outside of the body. The majority of these products will be applied as creams, ointments and oils to the skin. Also contained within the category of topicals are eye preparations which are aqueous solutions or ointments. Finally, it is convenient to put implants, which are placed under the skin, in this category.

Because the product is applied to the skin the need for sterility is less than that for injectables. However, this must be taken as a generalization, as eye solutions and some creams, ointments and oils must be sterile.

Non-sterile products do not require the best environmental conditions. Rooms which are Class 100 000 would usually be acceptable but the preparation of eye and other sterile products will, depending on their method of production, require an environment similar to that used for the production of aseptic or terminally sterilized injectables.

Implants have special contamination control problems. Because they are placed into the body they should be produced sterile but because of their very high biological activity they must be produced in containment areas to prevent them reaching the employees in significant concentrations. This may be achieved, for example, in 'Class II' or 'Class III' types of workstations (see Chapter 1) or in containment isolators.

Oral Products

Into this category can be placed products that are drunk or swallowed and also pessaries and suppositories. These products include bottles containing liquids as well as tablets and capsules. Oral products are not normally sterilized but rely on the quality of the raw materials and, in the case of aqueous liquids and creams, preservatives are frequently added to prevent development of any bacterial contamination in the container. Most of these products are also produced in closed systems where the raw material goes in one end and the finished product comes out the other end. The problem with these products, especially tablets and capsules containing powders, is one of cross contamination between products or between products and people.

Whilst of interest in the overall scenario of pharmaceutical production, they do not normally require closely controlled cleanrooms.

FACILITY DESIGN

Design Objectives

The design objectives of a pharmaceutical cleanroom suite, within a manufacturing facility, can be summarized as follows:

- Exclusion of the environment external to the suite of cleanrooms
- Removal or dilution of contamination arising from the manufacturing process
- Removal or dilution of contamination arising from personnel working in the area
- Containment of hazards arising from the product
- Control of product-to-product cross-contamination
- Protection of personnel
- Control and management of the flow of material through the process steps by means of layout and configuration
- Control and management of personnel movement by optimizing the arrangement and connection of individual rooms
- Overall security of the operation by control of the entry and egress of personnel and materials
- Optimum comfort conditions for personnel
- Special environmental conditions for products, e.g. low RH for powder filling
- Accommodation of process plant and equipment to ensure safe and easy use, as well as good access for maintenance
- Effective monitoring of the conditions of the room

All these functions are important, and a suite of cleanrooms should be designed so that they overlap and function well. The latter parts of this chapter will develop these objectives in more detail.

Use of Guides and Standards

It is clear from Chapter 2 of this book that there is no single definitive standard or guide which can be applied to designing and operating a pharmaceutical clean area. The designer and user has available an array of cleanroom specification documents such as Federal Standard 209 and various national Guides to GMP as well as guides on isolator design (see Chapter 2). It is therefore essential that the designer develops a clear technical and performance specification with the user which will describe the design, validation and operation of a facility and process such that the product and regulatory demands are met. The appropriate regulatory authorities should be involved at an early stage as a step towards achieving their essential ultimate endorsement of the facility.

It is the writer's view that the available cleanroom standards broadly provide such an effective framework, so long as they are not taken as being absolute. The problem of treating such standards as absolute will become clear as the various cleanroom functions and environmental parameters are described.

Design Methodology

As a pharmaceutical processing suite has more roles than purely environmental cleanliness, it is necessary to analyse all the requirements and develop the solution in an organized manner.

A simplified stepwise approach can be summarized as follows:

- Analyse production stages
- Prepare process flow diagrams
- Define activities associated with rooms
- Define environmental quality requirements
- Quantify production, process and space requirements
- Prepare room association diagrams
- Define the accommodation needs
- Develop layouts and schemes
- Prepare designs and specification
- Undertake the detailed design and construction process

Depending on the physical size, scale and complexity of a facility, these steps can be carried out at various levels of detail. It is important to clearly define the responsibilities of all the parties involved, and ensure that experience is appropriate in those responsible. The material generated during the design process can be used in the discussions with the regulatory authority, and to assist a designer not familiar with the particular problems of a certain factory, to effectively understand the philosophies, principles and conditions being developed for the new facility.

Cleanroom Suite Layouts

The individual cleanrooms assembled to form a suite of rooms for pharmaceutical manufacturing will have clearly defined functions. These are illustrated in Figures 4.1 and 4.2.

Figure 4.1 shows a typical suite of cleanrooms configured to meet the requirements of producing an injectable product that can be terminally sterilized. Production personnel would enter the suite of cleanrooms through the 'clean changing area'. In this room factory clothes are removed, hands washed, and appropriate cleanroom garments donned. Raw materials and components, such as containers, would enter through their respective entry airlocks. In these airlocks procedures are used to reduce

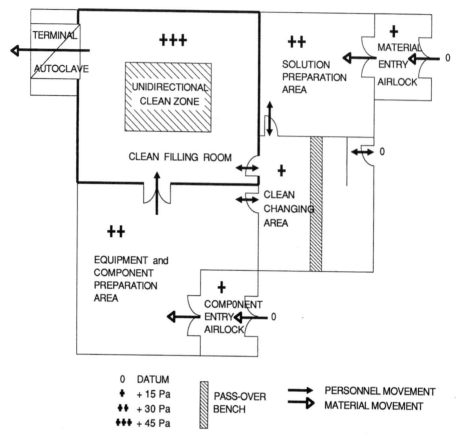

FIGURE 4.1. Typical suite of rooms for terminally sterilized injectables.

the contamination which may be introduced from outside the cleanrooms. Solutions are prepared in the 'solution preparation' room for transfer, directly or indirectly, by pipes or mobile containers, to the filling operation in the 'clean filling' room. Primary containers and closures would be prepared and washed in the 'component preparation' room and manually transferred to the filling stage or by using a conveyor system. Containers are filled and sealed under the unidirectional flow clean zone in the 'clean filling' room. Once filled and sealed, containers of product leave the cleanroom suite via the terminal sterilization autoclave. Upon completion of a work period, personnel would leave the suite via the changing room where cleanroom garments would be removed.

Figure 4.2 shows a typical suite of cleanrooms configured for the production of a product employing an aseptic filling technique. The differences in the process requirements lead to the following key variations: the complete suite is segregated into a clean suite and an aseptic suite. The barriers between the two are created by the oven, autoclave and transfer hatch for items entering the aseptic suite, and through the separation of the 'solution preparation' and 'aseptic filling' rooms. As the overall environmental control of the clean and aseptic suites is different, separate and more stringent

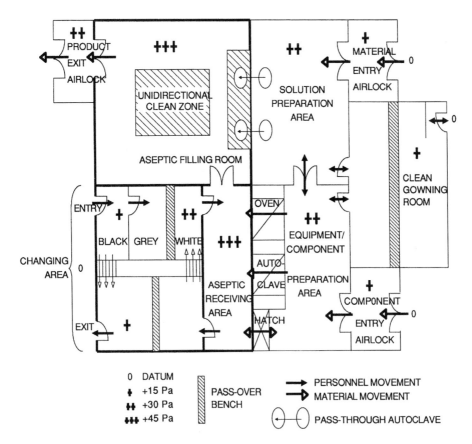

FIGURE 4.2. Typical suite of rooms for aseptic filling.

changing room control is provided for the aseptic suite. It is also possible that an isolator may be used in place of the unidirectional flow workstation.

ENVIRONMENTAL CLEANLINESS

The correct quality of cleanliness of the internal environment is usually the most difficult requirement to achieve and is determined by:

- The amount of contamination released in the room.
- The quality of the air supplied to the room.
- The quantity and method of supply of room air, i.e. conventional/turbulent ventilation or unidirectional flow, or a combination of both.
- The amount of ingress of contamination from areas adjacent to the room. When isolators are used, many of the same considerations are required, but generally the ingress of contamination from outside the isolated volume is minimized.

These requirements are considered overleaf:

Contamination Generation and Release

Contamination may come from:

- Personnel
- Process equipment
- Surfaces

The most important contaminant in a pharmaceutical cleanroom is bacteria and most, if not all of these, come from the people in the room. It is therefore useful to know the number of people expected to work in the rooms as this will have a direct bearing on the quantity of air required to dilute and remove the airborne dispersion of contamination from their bodies. The effectiveness of their cleanroom clothing will have a direct bearing on the contamination dispersed by the people in the room and hence the air quantity. This is discussed in more depth in Chapter 1. The type of clothing will also influence the cooling load, as the more effective the clothing is in preventing dispersion, the less exchange of air there is through the clothing fabric. Personnel will therefore be hotter and likely to require lower room temperatures.

Process equipment is a significant source of particulate contamination. Prevention by removal of particles at source should be the first objective before an allowance is made for removing it once it has entered the cleanroom space. This will ensure a more cost-effective design.

Cleanroom surfaces can be sources of contamination due to surface shedding caused by lack of cleanability and durability. The type of surfaces for use in cleanrooms are discussed in Chapter 9 of this book.

Ingress Through Defective HEPA Filter Systems

Within both unidirectional devices and room air supply terminals, similar HEPA filter cells will be employed. These filters are unlikely to have as high a particle removal efficiency as those commonly used in the most demanding parts of the micro-electronics industry. The reason for this is threefold. Firstly, bacteria do not normally occur in the air as unicellular organisms but are rafted on material such as skin particles and have an average size in the air of about 10 to 15 μm. Secondly, small inanimate particles because of their low deposition rate have little chance of gaining access into containers from the air in sufficient numbers to be a problem to patients. Finally, very small particles are not known to cause harm to patients. HEPA filters with an efficiency of about 99.97% to particle sizes around 0.5 μm are more than adequate.

HEPA filters must pass, after manufacture, an approved test in the factory. However good a filter element is at the time of manufacture, its final effectiveness will be dependent upon the care with which it is handled and the quality of the installation hardware. Effective housings or enclosures for the filter must be provided, and when the filter is fitted into the housing the whole system tested by an *in situ* test, such as the cold- or hot-generated oil test or other approved artificial aerosol test. A proven high efficiency filter and frame produced by a reputable manufacturer is most likely to ensure a satisfactory conclusion.

By-passing of contaminated air due to a poorly constructed filter housing is common. The application of the DIN 1946 pressure-seal test is a valuable provision within terminal filter housings, as is the use of fluid seals which have proved to be more tolerant of constructional inaccuracies, and avoid the need for high compression forces to ensure effective gasket seals. Further discussion of HEPA filters and a description of these filter frame housings is given in Chapter 8.

Another important element in correct terminal filter installation is ensuring that the terminal filter housing into which the filter is placed is connected effectively to the ceiling membrane of the cleanroom. The achievement of a good working detail at this point of connection will prevent induction of contaminated air from the void space and also avoid particle release from building construction materials that might otherwise be exposed at the interface between the filter housing and the building construction surface material.

Contamination Removal in a Room by Displacement or Dilution Ventilation

The accepted options for ventilating a complete room to remove undesirable contamination are either unidirectional flow systems or turbulent air flow, or a mixture of both. As discussed above, it is common in pharmaceutical cleanrooms to find that in a mixed flow solution the general room area is ventilated by a conventional turbulent system and unidirectional flow units are used to protect the areas where the product is exposed directly to contamination, i.e. the container is open. A typical cleanroom is shown in Figure 4.3.

This type of mixed solution is prompted by consideration of:

- Economy, as less air is required.
- The complex nature of pharmaceutical cleanroom suites which makes it difficult to return the large air volumes required in a full unidirectional system. This is particularly difficult if perforated floors cannot be used because of the chance of glass or fluids falling onto the floor.
- Relative room-pressure control requirements. Unidirectional flow rooms are normally supplied by individual air conditioning plants. This can make the pressure balance between rooms more difficult to achieve.
- Segregation. Users often require that the critical manufacturing operations are clearly seen to be segregated from the rest of the room. This is easily done by use of a localized unidirectional flow system.

If the amount of contamination liberated into the air of a cleanroom is known, it is possible, in a conventional flow system, to estimate the air supply volume required and hence obtain the required standard of airborne contamination. However, information of the dispersal rate of staff and, especially, machinery is difficult to obtain. It is therefore normal to use the air quantities given in Standards or Guides to Good Manufacturing Practices or to base these on experience.

It should also be recognized that the type of air supply devices, i.e. grilles or diffusers, will influence the air movement in a conventionally ventilated room and hence

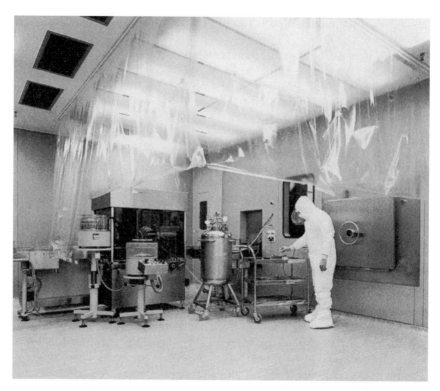

FIGURE 4.3. Vertical flow clean zone over autoclave discharge and vial filling machine. General room environment controlled by conventional turbulent air flow system with low level return air. Curtains pulled up to show equipment.

its cleanliness. Many types of air supply devices are available but on the basis of their method of action they can be divided into two types. These are:

- Eggcrate or perforated-plate type giving a downward jet of air flow.
- Various types of bladed or swirl diffusers which give good room air induction.

These two types of air supply devices give an air movement of the type shown in Figures 4.4 and 4.5. The eggcrate, perforated or dump-type of device will produce a jet flow of air below it. This jet of air will entrain contamination at its edges but, in general, the quality of air below the diffuser will be good. The bladed type of device is designed to entrain room air and mix it with the incoming supply air; the quality of the air throughout the room will therefore be reasonably constant. it may therefore appear that it is best to fit dump-type air devices and to place the critical operations under them. However, conditions away from this area must be poorer and it is not unusual to find that critical operations are resited away from the designed position. It is therefore the belief of the authors that air supply devices which give the best mixing and hence constant airborne contamination throughout the room should be used with unidirectional devices for enhanced local air conditions.

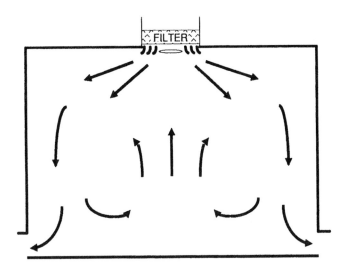

FIGURE 4.4. Air movement induced by bladed diffuser.

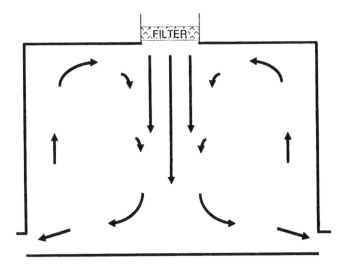

FIGURE 4.5. Air movement induced by 'dump' air supply.

The calculation of the air supply volume requires knowledge of the following:

- Minimum air change rates to meet the appropriate cleanroom standards for removal of internal contamination.
- Internal heat gain.
- Air volume required to pressurize the suite.

Experience indicates that a modern manufacturing facility, used for the production of injectables, will contain process equipment that liberates heat into the space which will require 25 to 35 air changes per hour for cooling; this is significantly greater than the minimum levels quoted in the cleanroom standards and very much greater than that required to pressurize the room.

Rooms should be designed so that the manufacturing process and the environment can be effectively monitored. To achieve the aims of the Guides t GMP it is important that there is no variation in the environmental conditions which could give unsatisfactory conditions and hence lead to contamination of the product. It is therefore necessary that the user is able to monitor conditions in the room. These conditions should therefore be defined at room validation and maintained routinely during manufacturing.

It is prudent to provide the more complex facility with a range of monitors to assist with fault finding. For example, air flow measurement and pressure switches provide valuable information as to system performance. Failures of such items as fans, filters and controls, etc. will change the room pressure.

The provision of unidirectional flow devices would normally be left to a specializes supplier. However, when specifying unidirectional flow devices or cabinets, it is essential either to write a specification, or apply accepted clean air device standards (see Chapter 2). The most frequent problems encountered with unidirectional flow clean air devices are filter seal penetrations, cabinet carcass leakage, airflow imbalance (in Class II types), and a failure to incorporate facilities for convenient testing of the integrity of filtration systems as well as performance monitoring.

Isolator and Barrier Devices

In recent years, isolator and barrier technology has been applied successfully to pharmaceutical manufacturing. Historically, isolators were used in the nuclear industry to manipulate and store radioactive isotopes and hence protect the operators. They were also used in veterinary research where they were use to keep animals germ-free. In pharmaceutical manufacturing they were initially used to provide clean conditions in the microbiological laboratory, where the testing of the final product for sterility was carried out. These clean conditions had to be at least as good, if not better, than were required in the manufacturing area to ensure that 'false positives', i.e. bacterial contamination arising in the laboratory, did not cause the rejection of a batch. Isolators are now used in many areas of pharmaceutical manufacturing. They can be used to protect either the product from contamination, or the personnel from potent compounds, or in some applications, such as parenteral cytotoxic products, they can protect both. The developement of a multiplicity of applications has meant that the original relatively simple design of an isolator has evolved and these evolved isolators are often referred to as barrier technology.

The use of barrier technology has grown largely from the requirement to increase both the protection of personnel and the product, but there may also be economic advantages gained through the reduction of the cost of providing and maintaining the manufacturing environment. In the case of the aseptic filling of pharmaceuticals, it is common for the total facility and process costs for an isolator application to be 70% to 90% of that of a traditional cleanroom system. However, greater cost savings can be made in running costs where a reduction of 40% to 50% can sometimes be achieved. In such cases the savings can come by:

- a reduction in the quantity of air needed for the maintenance of cleanliness or asepsis;
- a reduction in the amount of sub-division of the space used in manufacturing;
- a reduction in the garment quality and the number of times a clean, processed garment is required;
- a reduction in the amount of monitoring of particles and microorganisms;
- a reduction in the area of critical surfaces requiring cleaning and disinfection;
- an increase in flexibility in the deployment of personnel;
- improved equipment utilization.

Conventional filling of aseptic products is normally carried out under a unidirectional flow workstation in a non-unidirectional ventilated room. In this situation it is normal for personnel to enter and work within the unidirectional workstation and they must therefore wear high-quality cleanroom garments to ensure that none of their bacteria reaches the product. It is also necessary to ensure that any air entrained into the workstation from the surrounding room will have a low concentration of bacteria. An isolator can exclude the bacteria from both the personnel and the surrounding room and hence allow a reduction in the quality of the garments and the room air. Because of the effectiveness of isolators in excluding bacteria, the cleanliness standards can be relaxed not only in the area surrounding the isolator, but in the air movement control measures used to minimize the air movement between the production room and its surrounding areas. This relaxation of the air movement control can, in turn, lead to a more simplified design of the cleanroom suite and its air-handling systems.

Under some circumstances it is possible to reduce the time required to build a new manufacturing facility using isolators. However, greater time savings can be made when upgrading a facility by the addition of an isolator, instead of upgrading the whole manufacturing facility.

Typical Isolator Designs and Applications   The formulation, filling and quality control of sterile pharmaceutical products can employ flexible or rigid isolator systems which can be used as linked networks or separate modules. The range of size and complexity of isolators is large and continuously growing as applications to specific processes are developed. Many of the more complex systems must be designed and engineered as process engineering projects, rather than standard equipment procurement tasks, to ensure that the particular requirements are adequately defined, designed, fabricated and assembled, tested, and set into operation. The design of isolators is principally influenced by:

- whether the requirement is to protect the product, or the personnel, or both; this will determine whether the isolators should be positively, or negatively pressurized, with respect to the room it is in;
- the type of transfer system required for the entry and exit of materials into and out of the isolator;
- the technique used to undertake manipulations inside the isolator;
- whether rigid or flexible walls should be used.

The maximum protection of the product, or process, inside an isolator will be achieved by using a positively pressurized isolator. Internal pressures in the order of 20 to 70 Pa, compared to the surroundings, are typical. Particular care must be taken to ensure that over-pressurization is maintained when gloves or half-suits are in use, as when leakage occurs it is often traced to this source. Where hazard containment is required, a negative pressure system would be selected. Where an isolator is required to give both combined product protection and containment, a decision must be made about the pressurization required. Where asepsis is to be maintained, a positive pressure would normally be selected and particular care taken in the selection of the transfer system.

The type of transfer device selected for the transfer of items into and out of an operational isolator will have a major influence on the capability of that system to maintain a defined level of internal cleanliness, or to contain a specified hazard. Typical techniques used range from the least effective air-flow-protected 'mouse holes' and transfer tunnels, to the very effective alpha/beta docking port devices.

Most isolator applications require some human manipulation capability. Other than when robots or remote manipulators are used, this capability is achieved by the use of glove ports or half-suits. These two techniques have a major influence on the isolator design since they have radically different ergonomic attributes and require completely different designs of isolator to accommodate them.

It is clear that rigid and flexible wall devices are basically different in terms of their durability, configuration and fabrication techniques. Generally speaking, flexible film isolators are simpler shapes, and cannot easily house complex air treatment system fans and air duct passages.

To further emphasize and review some of the objectives identified above, the designs utilized in different applications are discussed and shown below.

Sterility testing. For many years barrier technology has been a valuable tool for providing clean conditions in the bacteriology laboratory for testing the sterility of the end-product. For this application, both flexible film and rigid wall devices have been successfully utilized in cleanroom and non-cleanroom conditions in the microbiological laboratory. These isolators are used to carry out manipulations with a bacteriological medium. Any failures in the security of the isolator will show up as growth in the bacteriological medium and so the effectiveness of barrier technology, operating in conjunction with sanitization techniques and transfer systems such as interlocking transfer ports, has been demonstrated. They can therefore be confidently applied to high-quality aseptic processing with the expectation of consistently high aseptic quality.

Sterility testing requires a strict control of the microbial contamination challenge from outside the controlled environment, but not of the particulate contamination liberated by the process itself. Hence, a positive pressure isolator in a controlled environment, using non-unidirectional airflow, is satisfactory. Flexible film devices using half-suit manipulation techniques are frequently used in these applications.

Shown in Figure 4.6 is an example of an isolator used in sterility testing.

Sub-division and dispensing of potent compounds. Since pharmaceutical products contain more and more potent actives, and as health and safety and environmental

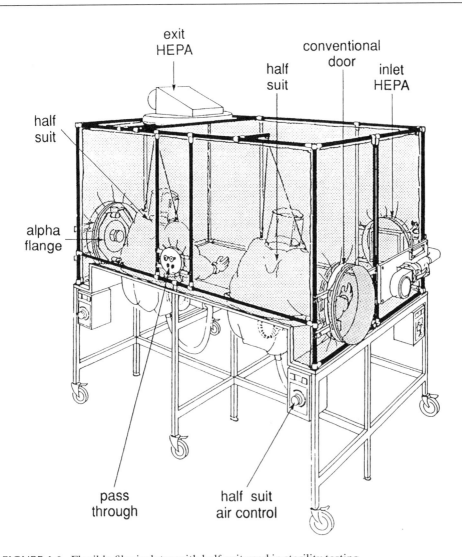

exit
HEPA

conventional
door

half
suit

inlet
HEPA

half
suit

alpha
flange

pass
through

half suit
air control

FIGURE 4.6. Flexible film isolator with half-suit used in sterility testing.

protection issues gain in stature, isolators have been developed in many shapes and forms to enable the safe weighing and sub-division of highly active compounds. The most sophisticated applications, such as the sub-division of the bulk sterile active compounds, require that the isolator maintains aseptic processing conditions internally. Figure 4.7 shows an isolator device configured to allow a keg of potent raw material to be introduced into an isolator environment, and to be securely sub-divided into lots suitable for a formulation batch process.

Powder handling systems. A natural extension of the handling of potent compounds is the use of barrier technology to provide highly secure mechanisms for processing and transferring powders in lots of, typically, up to about 50 kg. Barrier technology appli-

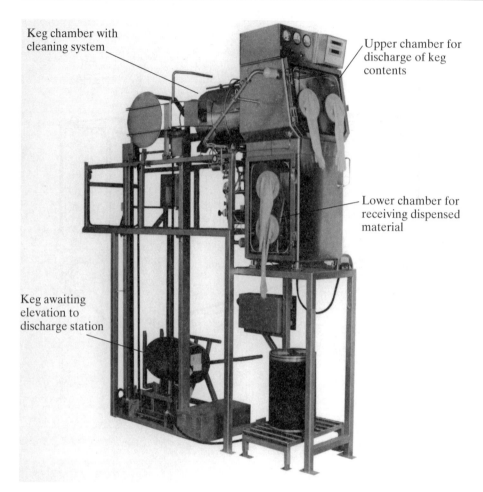

Keg chamber with cleaning system

Upper chamber for discharge of keg contents

Lower chamber for receiving dispensed material

Keg awaiting elevation to discharge station

FIGURE 4.7. Potent drug dispensing isolator.

cations in this area have the two objectives of operator and environmental protection, as well as the provision of a clean or aseptic environment around the process.

Shown in Figure 4.8 is an example of a system used for the drying and handling of bulk powder. As well as providing an aseptic environment for handling bulk powder it also provides for the containment of the potent compound and a nitrogen environment to allow the use of flammable solvents. It is a rigid positive pressure device located in an EC Grade D cleanroom (this is approximately equivalent to ISO 8 or Class 100 000). The cleanliness of the internal nitrogen environment is maintained using non-unidirectional airflow. Manipulations are achieved using glove and half-suit systems.

Small-scale manipulations. Many isolator applications at the clinical trial scale of manufacturing are based on the scale of technology used in sterility testing. The aseptic dispensing of pharmaceutical products in hospital pharmacies is also carried out on this scale. Manufacturing is not carried out on a continuous basis but in batches. Because of

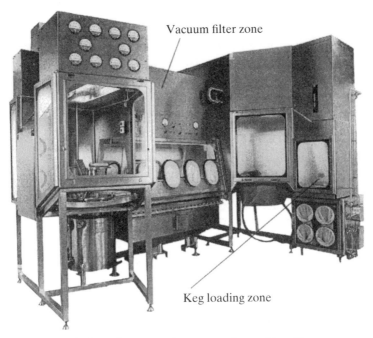

Vacuum filter zone

Keg loading zone

FIGURE 4.8. Specialized isolator for potent drug processing and handling.

this, the batches can be transferred out of the isolator taking advantage of one of the more secure transfer systems. In this type of application it is common to find either one isolator being used to dispense a variety of products, or more than one isolator used for separate tasks.

In the application shown in Figure 4.9 a family of isolators is placed in an EC Grade C working environment (approximately equivalent to ISO 7 or Class 10 000) and are configured to provide an aseptic environment interfacing with a depyrogenating oven

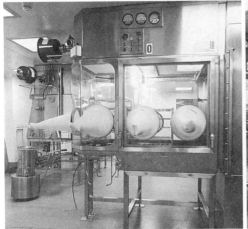

FIGURE 4.9. A group of isolators used in clinical trials and specials manufacturing.

for handling and holding dry heat sterilized components (left hand picture). Alongside this isolator, further individual isolators are provided in which separate formulation and/or filling operations can be undertaken (right hand picture). Materials are moved from one isolator to another using closed containers that dock with alpha/beta docking ports mounted in the side wall and floor of the isolators. Internal cleanliness is achieved through positive pressurization and the supply of double HEPA filtered unidirectional airflow to the critical process zones within the isolators.

Large-scale aseptic production. Isolators are now used in industrial scale aseptic processing for both formulation and filling. Shown in Figure 4.10 is one such example. In this example there is a rigid interconnected network of isolators providing a complete component handling and aseptic filling capability in an integrated line. The internal controlled environment is achieved using a combination of non-unidirectional and unidirectional airflow within positively pressurized cells. The individual cells are separated by 'airlock' flap valves. The product flow through to aseptic filling is largely automatic, with gloves provided for specific manual interventions. The complete line is placed in a room in which the environment is controlled but unclassified in terms of cleanroom standards, and utilizes hydrogen peroxide vapour for both the surface sterilization of the isolator network, and for the continuous surface disinfection of containers of product components that enter the system.

Transfer Methods   There is a variety of methods used to transfer material into and out of isolators. These are as follows.

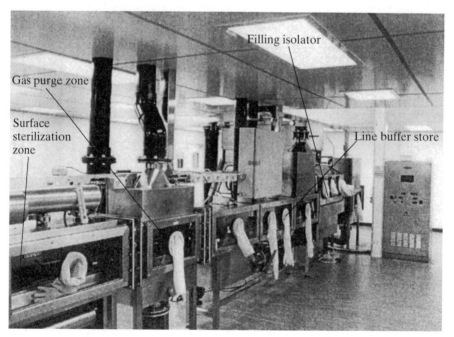

FIGURE 4.10. A specialized network of isolators used for large-scale industrial syringe filling. Process flow is from left to right.

Solutions/liquids. When product solutions are brought into an isolator environment it is critical to evaluate carefully the method of transfer and the connection and disconnection of the transfer system. The most secure approach to adopt is a closed piping system using clean and sterilize-in-place techniques. Where this is not practicable, the use of other aseptic connection techniques should be considered. Specially adapted docking ports, split butterfly valves, and sanitized connection techniques should be evaluated for the particular application.

Materials. The most secure method of transfer of material is that of transfer ports described variously as Rapid Transfer Ports (RTPs), High Containment Transfer (HCT) ports, and Alpha/Beta Flange ports. Their purpose is to allow two separate contamination-free enclosed spaces to be connected and disconnected whilst maintaining separation from the outside world. This is achieved by a rotational docking system which can be used (i) to connect a small container to an isolator to transfer materials into and out of it or (ii) to connect one isolator to another. Figure 4.11 illustrates the principal of operation of a typical device.

Step 1
Container (or transport isolator) approaches closed isolator port.

Step 2
Container docks with port and rotates to lock and enclose exposed faces. At the same time the interlock is release on the isolator door.

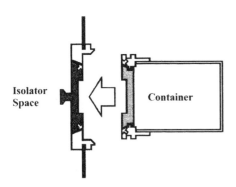

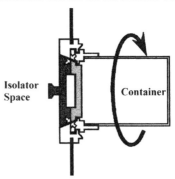

Step 3
Door into isolator is opened to allow free communication between the two enclosed volumes.

Step 4
De-docking is the reverse procedure.

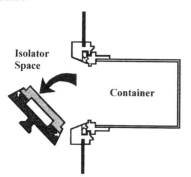

FIGURE 4.11. A rotational docking system.

The docking port method described above is the most secure but it is an expensive method of manipulating materials into and out of an isolator. Other simpler but less secure methods are available. Devices such as 'jam pot' covers, or single doors, present very little ability to separate the external from the internal environment. In fact, in these applications the only facet of the device's performance that will provide any protection is outward airflow when the door or cover is open. Generally speaking, these devices should not be opened once an isolator has been set up for use and only opened when the process is complete. An improved level of security can be achieved with a double door pass-through hatch, and the performance and effectiveness of such a device can be progressively improved by introducing mechanical or electro-mechanical interlocking of the opposing doors and, furthermore, by introducing positive ventilation of the airlock space to dilute and remove contamination that may enter when the external door is open.

The configuration of 'transfer hatches' or 'pass boxes' (Figure 4.12) can be divided into the following basic types with a progressively greater ability to segregate the internal from the external environment:

- *Unventilated transfer hatches with no interlocks on doors.* This solution is the least secure because the transfer hatches allows contaminated air to pass from the

FIGURE 4.12. The typical transfer hatch or pass box shown in this figure can be provided with an appropriate combination of interlock and ventilation attributes of the type explained in the text.

outside to the inside and vice versa, and also allows the internal isolator area to be open to the outside should the control procedures fail.

- *Unventilated transfer hatches with interlocks on doors.* This approach is similar to the unventilated transfer hatches, but has the added security of an interlock to prevent both doors being opened simultaneously.

- *Ventilated transfer hatches without door interlocks.* This approach adds ventilation to flush away the potential contamination trapped in the transfer hatches. It relies upon control procedures to avoid simultaneous door opening, and to acheive a defined flushing period.

- *Ventilated transfer hatches with door interlocks.* This is the most secure transfer hatch arrangement. This concept can also be provided with a time delay to allow the chamber to be purged effectively before the opposing door is opened.

The method of transfer using a transfer hatch or pass box is generally used where the manufacturing process, or product testing, is carried out in batches. However, when the process is continuous, as with large-scale manufacturing, then the product can be continually transferred out of the isolator either by (i) using a final holding isolator and using one of the methods described above or (ii) through a 'mouse hole' or transfer tunnel. Such later methods require careful qualification and performance monitoring at start-up and during their operational life. An airborne challenge generated in the surrounding room and measurement of the penetration of the challenge into the isolator is an appropriate test of the segregation capability of such arrangements. The effectiveness of the transfer system selected will be influenced by the standard of the surrounding room and vice versa (see discussion below).

The Environment Surrounding Isolators   The quality of the environment surrounding an individual isolator, or network of isolators, must be determined. The decision must take into account a number of things. These are:

- process risk;
- barrier integrity;
- manipulation techniques;
- transfer techniques;
- internal pressurization;
- sanitization techniques;
- regulatory requirements.

The process risk. The susceptibility of the product to contamination is an important consideration in terms of risk analysis. The first consideration is whether the product is to be produced aseptically or to be terminally sterilized. Aseptic products, particularly those which could support the growth of bacteria, are likely to benefit from the use of an isolator system. Terminally sterilized products are unlikely to benefit, unless there is a particular problem with particle contamination from the room. The likelihood of microorganisms entering and growing within the container is dependent on (a) whether the containers are delivered to the filling area are the open or closed type; (b) the area of the neck; (c) the time it is open; and (d) the chance of bacteria growing in the product. An evaluation of all steps of a process from the aseptic assembly of the system through to the product transfer will give a risk assessment.

Barrier integrity. An isolator should be manufactured in such a way that the barrier integrity can be maintained, and demonstrated to be maintained, over its lifetime. It is clear that the greater the barrier integrity, the less opportunity there is of bacteria entering it from outside and it follows that the surrounding environment will have a lesser impact on the process or activity contained. Considerations of rigid compared with softwall envelopes, the quantity and quality of access panels (these always present a risk when resealing, etc.), the type and nature of joints and sealants used, the quality of fabrication, and thoroughness of fabrication tests are very important. It is most important to be able to routinely carry out barrier integrity tests by pressure hold, decay rate, or leakage rate tests. The easier it is to effectively carry out such tests and demonstrate the effectiveness of the barrier, the less demand there will be to provide a high-quality background in the area where the isolator is used.

Manipulation technique. The manipulation technique (gloves and sleeve, gauntlets or half-suit) to be employed will influence the overall integrity of an isolator, since where the manipulation device is connected to the isolator there is the possibility of a breach to the barrier of the isolator. Additionally, where the manipulation involves placing part of the human body within a device, such as gloves or a half-suit, a greater potential exists for process or product contamination than would occur with tongs, remote manipulators or robotic devices. It is the authors' opinion that where human interface is required, glove and sleeve or gauntlet applications are preferable to the use of half-suits. This becomes particularly important where routine tests are used for determining *in-situ* the integrity of the manipulation device. The principal of utilizing a simple pressure hold or gas diffusion leak test on a glove port and glove before and after a process work session can be used to achieve a high level of glove integrity assurance. Such a technique is very difficult with a half-suit application due mainly to its relative size. Alternative methods of tongs and remote manipulators (derived principally for handling radioactive substances) can also be of advantage in isolators which are utilized for the maintenance of asepsis or containment of toxic and biological hazards. However, it should be noted that tong manipulators require a gaiter system to ensure air pressure integrity around the rotation and sliding gimbal. This gaiter will be akin to the glove gauntlet in its integrity and risk of failure.

Transfer technique. The transfer of materials into and out of an isolator represents another route of loss of internal environmental integrity. Clearly, the more secure the transfer system the less demanding the clean conditions of the external environment become. For isolators producing small batches, the most secure techniques are inter-locked docking port systems (often called alfa/beta systems) and the least secure is 'jam pot' covers or single doors. For continuous production systems the best method is airflow protected tunnels, and the least effective are 'mouse holes', This is discussed further in the 'transfer methods' section above.

Whilst no standard test exists at present to define the performance of such transfer devices, containment tests such as those used for open-fronted microbiological safety cabinets can be effectively used to determine a protection factor . These quantify the segregation achieved by such devices in operation. This type of test is, in the authors' opinion, likely to become a basis of type, or performance, testing of transfer devices.

Internal pressurization. The internal pressure within an isolator clearly has a major influence on the ability of the isolator to exclude the external environment. The level of pressurization should also be considered in relation to its ability to withstand the piston effect of rapid glove movement, and whether or not the internal isolator air system should achieve an outflow of air in the event of partial or total glove loss. The glove piston effect is relative volume related (i.e. the volume displaced by the glove compared with the volume of the isolator). As a rule of thumb, devices with a pressure of 15 to 25 Pa compared with the surrounding atmosphere are likely to be less secure due to the piston effect than devices with a pressure of between 50 Pa and 80 Pa compared with the surrounding area. When airflow protection is required to ensure that there is sufficient air to compensate for the partial or total glove loss, then air velocities of between 0.5 and 0.7 m/s per second through the unprotected area should be considered. Such protecting velocities can be produced by designing the air handling system to ensure that it can intake sufficient room air in the event of glove loss and its associated internal pressurization loss. There is a distinct possibility that velocities much above these figures may cause re-entrainment of external contamination due to high turbulence.

Whilst this chapter is mainly focused upon positive pressure devices, the integrity and performance attributes of pressurization equally apply to negative pressure systems. However, negative pressure systems are designed to draw air into the isolator and therefore are likely to require a higher class of surrounding environment than that for equivalent positive pressure systems.

Sanitization methods. How the isolator is cleaned or sterilized, and its frequency, is important. The effectiveness and repeatability of the sanitization method has an impact upon the quality of the surrounding environment, particularly if it is anticipated that batch change-over is carried out with an isolator open to the room. If an isolator is placed in a room in which the environment is controlled but not classifiable as a cleanroom, it is necessary to minimize the introduction of room contamination into the open isolator. This can be done by operating its on-board air system in a total loss rather than a recirculation mode. Assuming that some contamination has entered the device, it is essential to use a highly repeatable and effective sanitization or surface sterilization method. Methods using surface swabbing and aerosol spraying with a disinfectant are unlikely to give a repeatable high efficacy of disinfection. However, the use of highly controlled gaseous-phase processes, such as traditional formaldehyde, hydrogen peroxide and peracetic acid, are very effective provided that they are applied to surfaces that are clean and free from soiling build-up. Aerosols of disinfectants can also be considered for some applications, but the issues of effective penetration and the attenuation due to soil build-up are more pronounced for this type of process.

Regulatory requirements. The decision as to what quality of room environment the isolator should be placed in can be determined by consideration of the above variables.

It is the responsibility of the isolator user, together with the designer or supplier, to demonstrate an appropriate level of system integrity to support the selection of the surrounding environment. Regulators will hope to see that the use of isolators will bring about a reduction in the contamination rate of the product as well as other facets of product quality assurance, operator health and safety, and environmental protection. Current experience is that where effective isolator systems can be created, and the integrity of the system continuously demonstrated as sound, the surrounding environment for aseptic processing within the isolator can be controlled but unclassified in terms of cleanroom standards. The most compelling evidence to support this statement comes from the field of germ-free animals, where batches of animals have been bred free of microorganisms for over 15 years in isolators kept in unclassified rooms. However, the drug regulatory authorities are nervous of this philosophy, and prefer the more conservative approach set out below.

Some knowledge of the control achieved will be required, in particular the particulate and microbiological challenge to the isolator system. This can be achieved by instigating a simple environmental monitoring programme in the room. There will also have to be clear evidence of control of access and entry of people into the operating space. The level of information and data collected and judgements made will vary from one regulatory authority to another. It is the authors' opinion and experience that it is appropriate to use a surrounding environment capable of achieving a particulate classification of ISO 8 (Class 100 000). More demanding surrounding conditions of ISO 7 (Class 10 000) should be anticipated where negative pressure or low technology positive pressure isolators are used. According to the EU GMP (1997) for aseptic processing, European regulatory authorities require the surrounding environment to be (at least) grade D. This is approximately equivalent to ISO 8 or Class 100 000.

Design Considerations for the Surrounding Environment   Having taken all the above issues into account and determined the type and nature of the barrier technology used and the quality of the surrounding environment, it is important to consider the broad range of issues relating to the quality of the surrounding environment. It is not sufficient to consider only whether the surrounding environment should be classified in terms of a cleanroom standard or an equivalent GMP microbiological requirement. It is necessary to consider, in a much more thorough way, all the attributes of the facility in which the isolator is placed, as this can have a major influence on the type of surrounding environment. In what follows we identify some of the key issues to illustrate that the subject is more than simply cleanroom cleanliness classification.

The size of the manufacturing space is clearly significant in relation to the physical cost of creating it, the energy to maintain the conditions, and the efforts to produce some appropriate level of monitoring of the quality of that environment. The smaller the space the lower these costs. However, experience with isolator applications generally shows that the total area of accommodation will not be significantly reduced by the use of isolators or barrier technology. Where an advantage can be obtained is in the reduced complexity and sub-division of the spaces. If a group of discrete isolators accommodating different steps of the process can be placed adjacent to each other within a single simple space, there are major advantages to be gained in terms of communication, people flexibility, circulation space sharing, and ventilation and air conditioning system sub-division. The design of a suite of rooms used in traditional aseptic production is shown in Figure 4.13 A design used with isolators could be simpler, and shown in Figure 4.14 is one such possible layout.

The quality of construction and materials used to create the facility should be considered, and a move from high-quality pharmaceutical cleanroom construction to good quality packaging department type construction can show significant cost benefits and construction time advantages. As mentioned earlier, simplified space and less demanding environmental requirements mean that HVAC systems have less demanding performance requirements, and can be far simpler in terms of the zonal control of air volume, temperature and room pressurization. These issues reduce both capital and operational costs.

Operational Issues/Revenue Costs   The other major advantage that can be obtained from placing an effective isolator application within the minimum necessary surrounding environment is the cost of maintaining and managing the surrounding environment. Reductions in effort, and hence associated cost, can be obtained in cleaning and disinfection as well as the monitoring of particulate and microbiological contamination. By removing the need to heat and cool vast quantities of air needed for the maintenance of cleanliness or asepsis to that required to satisfy the comfort of the occupants, significant energy savings can be achieved. Revenue costs associated with personnel control, particularly the changing of garments, are significantly reduced by

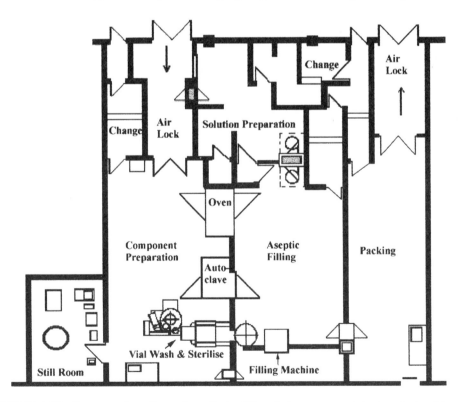

FIGURE 4.13. Conceptual configuration of a traditional cleanroom cytotoxic paterteral formulation and filling facility.

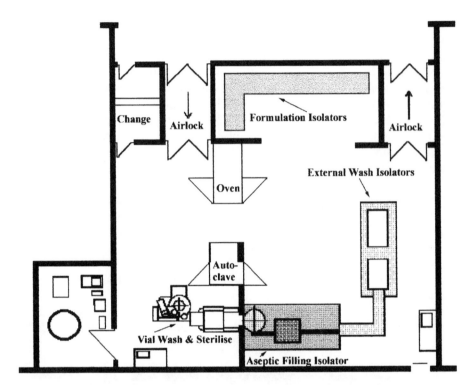

FIGURE 4.14. Conceptual configuration of a cytotoxic parenteral formulation and filling facility using isolators for the formulation, filling and external decontamination process steps.

both simplified space with reduced sub-division. The changing of garments for each entry is likely to be unnecessary and the quality of the cleanroom garment is significantly less than for the cleanrooms using traditional aseptic processing in cleanrooms.

Room Pressurization and Air Movement Control

The ingress of airborne contamination from outside the cleanroom is a frequent problem, although good design can limit this. This can occur when outside airborne contamination is induced through badly detailed fabric into the cleanroom. Gaps in fabrics should therefore be kept to a minimum and the room pressurized to minimize or prevent this problem. Pressurization of rooms is discussed below.

The ingress of contamination can also occur when personnel, equipment and material are transported through badly designed airlocks and changing areas; this may be surface or air contamination.

Pharmaceutical cleanroom suites consist of several cleanrooms within which different steps of the manufacturing process are undertaken. As the product materials and packaging components are taken through their processes and into different rooms, steadily increasing standards of environmental control are required until one reaches the stage of product filling and container closing and sealing (see Figures 4.1 and 4.2). Here the highest quality condition is required. As the sealed product goes on to labelling and inspection less stringent environmental conditions are required. These different standards of environmental control are achieved by different air supply rates and the use of unidirectional flow units or isolators at the critical areas.

To ensure that these different conditions are held in each cleanroom, Cleanroom Standards and Good Pharmaceutical Manufacture Practice guidelines require that rooms are maintained at different pressures. This is to prevent an undesirable flow of air from a lower grade area to a higher grade area, thereby reducing the possibility of contamination transfer. Experience, particularly with complex pharmaceutical processing suites involving many rooms, shows that achievement of a sensible relative room pressure regime and its subsequent maintenance presents some of the major design, commissioning and operational problems.

Within the established cleanroom standards there is a consensus view that the room pressure differential between cleanrooms should be 10–15 Pa. This is a level that is readily achievable, easy to monitor and appears to prevent contamination transfer. It is worth bearing in mind that although cleanroom guides may specify a differential pressure of 10 or 15 Pa this requirement is only a means to an end. If there is no adverse air flow between the rooms in the suite the pressure difference is not important. However, this argument may not be understood, and therefore not accepted, by the user or Regulatory Authorities. An exactly similar issue occurs in relation to isolators. However, in this case because the controlled environment is small, the displacement effect of gloves is important and must be taken into account when selecting and qualifying pressure differentials. Typical pressure differentials for isolators are 15–60 Pa.

In some situations the cleanroom's exhaust may be in an outside adjacent corridor reached through an airlock or change area and may be in an area two levels of pressure lower than the room. One must, therefore, bear in mind the limitations of the construction fabric in accommodating such static pressures as well as the capability of air handling systems in generating the required over-pressure. Excess differential pressure of over 30 Pa can cause 'whistling' through the door cracks and may give slight difficulties when manually opening and closing swing doors.

Process equipment that crosses room-pressure boundaries may cause a problem. This can be illustrated by considering a tunnel process where a component is washed, sterilized, and filled as it passes from a component preparation area into an aseptic filling room. The pressure differential will cause air to flow between the two areas joined by the tunnel. This air flow can change the heating characteristics, and hence efficiency, of the hot air oven and may even lead to hot spots and damage to the tunnel. Fluctuation in the pressure differential will cause changes in the air flow quantities which in turn may lead to changes in efficiency and difficulty in validating the system. Because the tunnel crosses boundaries between areas of different pressure it is necessary to constrain the flow through the tunnel by some type of masking. If this is not done, excessive air will be required to be supplied to the area of highest pressure. Even with a well-masked tunnel an appropriate allowance must be added to the supply air; this can be calculated from Equation (1) given later in this section.

Two methods by which cleanrooms within a suite may be pressurized with respect to each other are available and known as the 'closed door' and 'open door' solutions. The 'open door' approach has been developed and is of particular value where airlocks are inconvenient or impractical to provide, e.g. hospital operating theatres. An example of a 'closed door' solution as applied to a cleanroom suite is shown in Figure 4.15.

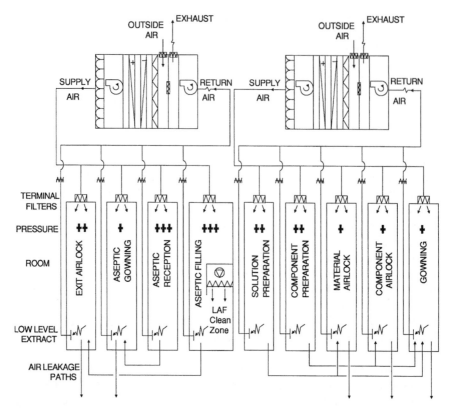

FIGURE 4.15. Air supply and extract in cleanroom suite using the 'closed door' approach.

The amount of air supplied to each room is that amount required to either meet the standards for dilution of contamination or for cooling requirements. The room's exhausts are adjusted to give the correct pressure differential. The air extract from the rooms can be adjusted manually or it is possible to fit automatic dampers which are regulated by the room pressure. The advantage of this type of closed door solution is its simplicity and the likelihood that it will work with few problems. As the air supply and extract in each room are almost balanced it will minimize the air transfer between areas.

It is necessary, as has been previously discussed, to ensure that the rooms are constructed in an air-tight manner to minimize the air flow out of the room's fabric. However, it is impossible to prevent air flow through the door cracks from an area of high pressure to one of low pressure. It is possible to calculate the amount of air leakage through small gaps and apertures by application of the formula.

$$Q = A \times \alpha \sqrt{\Delta P}, \tag{1}$$

where

Q = air volume (m³/s),
A = leakage area (m²),
ΔP = pressure difference (Pa), and
α = coefficient of discharge (0.85).

A good estimate of door leakage can be calculated if the tolerances of the door fit-ments are specified but the total leakage will depend on the quality of constructional detailing and this is unlikely to be finally known until the commissioning of the room is carried out. It is therefore wise to ensure the air handling system has adequate capacity to accommodate more leakage than is expected.

The disadvantage of the closed door solution is that it does not counteract the adverse air flows that occur when doors are open. Air flow will occur when a person opens and closes a door but if a door is left open there is a movement of air between the areas separated by the door which is caused by the turbulence of air generated by the throw of the air from the supply diffusers. This is greatly increased by the temperature differential between the two areas. For example, a double door will allow an air exchange of 0.19 m³/s in both directions when there is no temperature difference and 0.24 m³/s when the temperature difference is 2°C. To prevent this, sufficient air must pass through the open doorway in the direction of the less clean area to prevent adverse movement of air. The amount of air that was found to prevent any serious backflow of air across a single doorway (0.9 m × 2.05 m) and a double doorway (1.4 m × 2.05 m) is given in Figure 4.16. This is for a temperature difference of between 0°C and 5°C, appropriate volumes being chosen from the calculated temperature differ-ence. The volume required for different doorway areas can be obtained by propor-tioning these values.

Shown in Figure 4.17(a)–(c) are diagrams of the 'open door' solution applied to the layout given in Figure 4.1. Shown in Figure 4.17(a) is the expected air flow within the suite when the doors are all closed and in Figure 4.17(b) and (c) the air flow through the suite when the doors from the clean filling area re opened in turn. Owing to lack of space it is not possible to illustrate the effect of opening the remaining doors in the suite. However calculations demonstrate that the air flow in the suite should be in the correct direction, i.e. from the clean to less clean areas. This solution to the air move-ment control requirements has been calculated by Peter Robertson, formerly of Building Services Research Unit, University of Glasgow.

The air supply volume required to ensure that the air movement will always be in the correct direction throughout the suite is about 0.69 m³/s to the filling room and about 0.63 m³/s to the solution preparation room. These volumes are those required for air

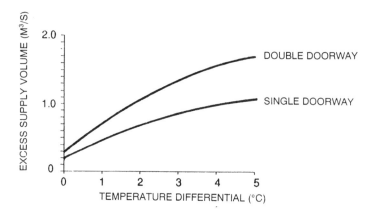

FIGURE 4.16. Air supply to isolate a doorway.

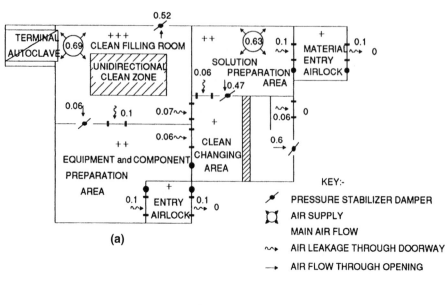

(a)

KEY:-

⚟	PRESSURE STABILIZER DAMPER
⊠	AIR SUPPLY
	MAIN AIR FLOW
⤳	AIR LEAKAGE THROUGH DOORWAY
→	AIR FLOW THROUGH OPENING

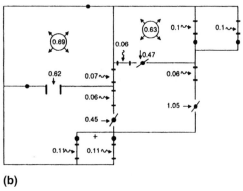

(b)

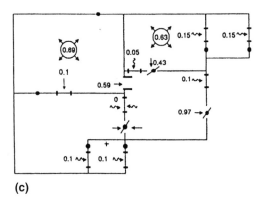

(c)

FIGURE 4.17. Air flow through suite: (a) doors closed; (b) door open between filling room and equipment and component preparation room; (c) door open between filling room and changing area.

movement control and take no consideration of the air requirement for the cooling load or that required to dilute the airborne contamination. They are therefore the minimum air quantities required for air movement control in a design of suite shown in the figure. Additional air can be used to increase the door protection, or can be extracted in the room it is supplied to.

These air volumes have been calculated for a maximum temperature difference of 2°C between the clean filling area and its adjacent areas and 1°C between all other areas. The grilles which allow continuous air flow when the doors are closed are pressure stabilizers and these should be set at the required room pressure and sized for the maximum air flow through them.

It should be noted that the values calculated for this design are unlikely to be exact as it is difficult to be certain of the room pressures when a door is open and hence be able to calculate the exact door leakage flows. However, the values will be sufficiently accurate for all practical purposes. An example of this uncertainty is the flow between the change area and the equipment and component preparation area when the door between the filling area and change area is opened. Because the pressure in the change area will not be exactly known there is the possibility of a slight reversal in flow into the equipment and component area. If this occurs it will be of little practical importance.

The disadvantage of the open door solution is that in those suites where there are many rooms and doors it is difficult to design a system that will ensure that there are no reversals of flow. However, if the airflow network is kept simple it is a viable solution.

TEMPERATURE AND HUMIDITY CONTROL

Target levels would be 20°C with an RH of 40%, ±5%. However, a lower RH will be required for moisture sensitive materials, e.g. powders; 25%, ±5% would be appropriate but individual products may have particular needs. Variations in dry bulb temperatures may also be required for economy or improved comfort conditions. Depending upon geographical location, the nature of the production and the clothing worn, it would not be unusual to see target dry bulb temperatures in the range of 18°C to 22°C.

LIGHTING LEVELS

Most pharmaceutical sterile products suites will rely upon permanent artificial lighting. The nature of the visual tasks usually requires a level in the range of 400 to 750 lux. Certain tasks may require higher levels, but these are usually provided by local lighting. At these high levels of illumination, colour rendering and glare become significant problems. The sealed and smooth nature of the cleanroom luminaires makes traditional glare control difficult, and this must be considered with respect to the level chosen. The light sources should be selected to have a colour rendering as close as practical to daylight.

NOISE LEVELS

Many of the manufacturing processes in pharmaceutical cleanrooms generate high noise levels and the high air supply volumes required will add to this. A level of approximately 65 dBA adjacent to unidirectional air flow devices has been indicated in some standards. This is too high and a design level of 55 dBA for the unoccupied cleanroom can be achieved in a well designed system. Clearly some tasks may demand lower levels, and the designer must evaluate specific needs.

AESTHETIC CONSIDERATIONS

The visual interest of a suite of cleanrooms is often forgotten. The occurrence of 'snow blindness' from high levels of illumination bearing on white surfaces has been recorded. Colour should be used in cleanrooms to prevent this and create a more pleasant environment. Colour can also be used to determine the boundaries of different cleanroom classifications.

The proportions of the cleanroom will influence the appearance of scale and space. If working to an air change rate, a minimum room volume is advantageous and this can be achieved with a low ceiling. However, this solution should be considered carefully when designing a very large cleanroom as this would not be aesthetically pleasing.

CONSTRUCTION, SERVICES AND EQUIPMENT

Cleanrooms are built using traditional building construction techniques, as well as specially developed prefabricated systems. Whatever system is used, it must give an effective and adequate compromise between something which is ideal in all respects and something which is buildable within the confines of cost and timescale.

From the preceding sections it should be clear that the quality of construction of the cleanroom suite has a direct bearing on the amount of air leakage out, or into, the room and such leakage should be prevented. It is also necessary to avoid ledges and crevices where contamination may gather, as well as ensuring that the quality of the surface finish is such that undue particle generation can be avoided. Windows and doors should be selected to avoid dirt traps and the junctions of finishes must also be satisfactory.

The air conditioning system, services and utilities, and process equipment must be integrated into the building fabric. The importance of ensuring that HEPA filter housings are integrated into the ceiling was discussed earlier and it is necessary that a similar level of detailed integration is necessary with respect to the room extract or return air terminal devices. Air ducts should also be air tight and cleanable. Other services and utilities similarly require a satisfactory sealed entry to a room, or to a piece of process equipment.

A feature often poorly considered is how production machinery can be repaired or replaced. Pharmaceutical cleanrooms and process equipment often require to be updated when changes are made to production activities. Most of these changes inevitably require rapid and convenient provision of additional electrical and mechanical services and may require changes to the suite's layout and construction. The provision of vertical service ducts in walls and service access over the cleanroom will allow changes to the services to be more easily accommodated. Physical layout changes will always be more difficult but the use of prefabricated walls may assist.

It is interesting to compare the arrangement of cleanrooms within the microelectronics industry to those in the pharmaceutical industry. Within the micro-electronics industry the servicing of cleanrooms is done readily through accessible service chases which interdigitize into the cleanroom (see Figure 3.1 in Chapter 3), and, by means of bulkhead fittings, services can be distributed to process equipment. This provides an easy method for subsequent servicing or modification of equipment without major disruption to the cleanroom space. However, this style of cleanroom is very difficult to achieve in the pharmaceutical setting as the sub-division of the cleanrooms tends to be very much greater and the need for connection of the different process steps found in separate cleanroom areas may be impossible if interposed by service chases. This clearly is a generalization and there are cases where service chases can be incorporated.

Lastly, virtually all project briefs will include the rather nebulous concept of 'flexibility'. However, requirements for such things as low level air extract, pressure differential maintenance and material and personnel movement management makes total flexibility an improbability. A philosophy of design is therefore best adopted which tends towards a facility where either individual suites, or cleanrooms within suites, can be modified without significant disruption to adjacent areas of the facility where production can still continue.

COMMISSIONING AND PERFORMANCE QUALIFICATION

Designing and building a pharmaceutical cleanroom is a complex task. Upon completion it is essential that the resultant cleanroom suite is shown to perform satisfactorily. This task will necessarily involve the designers, installers, and the users. For simplicity, this final task is best separated into commissioning and performance qualification. These tasks will merge together but for clarity it is best to consider these separately.

Commissioning

The commissioning process will require the installers to ensure that the environmental control system are complete and that the components are installed, operating correctly and achieving their technical design specification.

This will typically relate to items such as:

- Fans and pumps
- Controls and control devices
- Air flow quantities
- Room pressure balance
- Thermal performance

Operational Qualification

Knowing that the individual items of the air conditioning plant are performing correctly, it is then necessary to show that the correct conditions are achieved within the cleanrooms. These tests will be performed in the empty room but may also require the production equipment to be in place. A number of these tests are only carried out in cleanrooms and may require specialist firms to carry them out.

Tasks that are likely to be carried out are as follows:

- Room temperature and humidity challenge tests. The room should be artificially loaded to determine that the correct conditions can be achieved.
- Device calibration of all parametric monitoring equipment such as:
  —flow measurement devices
  —pressure transducers and switches
  —temperature sensors
- Pressure differential tests. The pressure differential between the areas should be checked.
- Air movement control tests. Each door in the suite should be opened in turn and smoke used to ensure that there are no reversals in airflow.
- HEPA filter integrity. These tests are discussed in Chapter 8.
- Airborne particle counts. The unoccupied room will be tested with a particle counter to establish that the number of particles in the room is not greater than the design specification. This will be done without the equipment working and, depending on the contractual requirements, with the production equipment working.

Tests which are less often carried out but still of use are:

- Room air pattern tests. These can be determined by visualization of the air by smoke release. These can be carried out in the room as a whole, but the critical area where the product is open should be studied. A video recording of the tests can be made. Air velocities and direction may also be measured.
- Particle decay rates. These will determine the 'clean up rate' at an important area. Smoke particles are seeded into the area under question and their decay rate ascertained by a particle counter. From this decay rate the actual effective air change rate in that area may be determined, or the time taken for the particle concentration to reach a percentage of the original concentration.

CONCLUDING REMARKS

The ultimate objective of building a cleanroom suite is to provide a vital element in the assurance of product quality within the overall philosophy of good pharmaceutical manufacturing practice. The resultant facility should prevent contamination of the product, and should be seen to be doing so by the incorporation of effective monitoring devices.

The facility and its operation must effectively control the contamination from people, raw materials, intermediate products, finished products as well as accommodating services, process plant and equipment. This is must do within the sphere of appropriate costs, complexity, and durability. The guidance available, and the requirements of those involved in the overall design and construction process are complex. It is therefore essential to ensure the whole process is undertaken in a controlled and organized fashion so that on completion the facility should be commissioned, qualified and validated to fulfil both the specification and the requirement for satisfactory production.

ACKNOWLEDGEMENTS

We wish to acknowledge the permission of Cyanamid Ltd to reproduce Figure 4.3, TPC Microflow to reproduce Figures 4.7, 4.8 and 4.9, La Cahene to reproduce Figures 4.6 and 4.11, Evans Medical to reproduce Figure 4.10 and Envair Ltd to reproduce Figure 4.12.

5 The Design of Cleanrooms for the Medical Device Industry

H. H. SCHICHT

INTRODUCTION

Medical devices encompass the broad variety of products used for medical purposes which do not achieve their action in or on the human body by pharmaceutical means (i.e. as medicinal drugs), by immunological means (i.e. as vaccines) or by metabolic means (i.e. as foodstuffs or beverages). Thus, they encompass mechanical or electronic devices and the accompanying software employed for diagnosis, prevention, monitoring, treatment and alleviation of disease, injury or handicap, as well as mechanical devices for conception control. Closely related to medical devices are surgical products such as wound dressings or threads.

To perform their functions, some products do not have to enter into the human body. Examples here are X-ray or electrocardiographic apparatus—two well-known tools employed in the diagnosis of illness or injury. Other devices, though, will have to penetrate into the human body to perform their functions, either into the natural body orifices (nose, ears, mouth, etc.) or through the skin. Of the devices or surgical products penetrating the human skin some, like syringes or surgical instruments, will interact with the human body for only a short period of time. Others, for instance orthopaedic implants or heart pacemakers, may remain in the human body forever.

From the point of view of contamination control, interest focuses on medical devices and surgical products which penetrate the skin or interact with open wounds: they are required to be sterile in order not to cause infections. A very high percentage of the total market volume of medical devices nowadays refers to such sterile products. Sterility, in absolute terms, means the total absence of microorganisms. In practical terms, this is usually interpreted as a sterility assurance level (SAL) of 10^{-6}—or one surviving microorganism in a million after sterilization. In the field of medical devices and surgical products, sterility is normally achieved through terminal sterilization, i.e. after the product has been packaged. Heat sterilization—commonly used in the pharmaceutical industry—is normally not feasible here, as the materials of which medical devices are composed do not, as a rule, withstand elevated temperatures. Instead, sterilization by the gaseous agent ethylene oxide (EtO), gamma radiation or electron beams are the methods most commonly employed.

Cleanroom Design. Edited by W. Whyte © 1999 John Wiley & Sons Ltd

THE CASE FOR CONTAMINATION CONTROL

For a device intended to penetrate the human skin, sterility alone is not enough: the product surface should also be free from particles as they might trigger undesirable effects in the human body. In addition, it should be free from dead microorganisms and from poisonous by-products of their metabolism, such as the fever-generating pyrogens. Moreover, in order to achieve the high safety level demanded from sterilization, a low bioburden to begin with—i.e. a low population of viable microorganisms on, or in, the product prior to sterilization—is an indispensable prerequisite.

This combination of requirements clearly establishes the case for cleanroom technology and contamination control practices during component preparation, product assembly, testing and packaging of sterile medical devices and surgery products. In fact, it can be assumed—as a rough estimate—that about 40% of the total world market volume of medical devices and surgery products are now manufactured under cleanroom conditions.

QUALITY SYSTEM PHILOSOPHY

As manufacturers of health products, the medical device and surgical products industry is highly regulated. A commendably systematic approach is apparent here in the philosophies for quality management and quality assurance. The point of departure is the ISO 9000 family of generally applicable International Standards for quality systems. For the European nations, the European Standard EN 46 001 provides a first layer of interpretation on how the ISO 9000 philosophy should be applied under the specific situation of the medical device industry. To be used in combination with the ISO 9000 series of standards, EN 46001 also embraces all the principles of Good Manufacturing Practices (GMP). Further and still more detailed additional guidance can be drawn from GMP guidelines such as the so-called *Blue Guide* issued by the British Department of Health with the title *Quality Systems for Sterile Medical Devices and Surgical Products—1990 Good Manufacturing Practice* (HMSO, London). Again structured consequently according to the ISO 9000 pattern, it reproduces in each chapter and subchapter first the relevant text from ISO 9001 in italics, followed then by the application-specific requirements and guidance.

In the United States the medical device is guided by the Federal Regulation 21 CFR part 820 *Good Manufacturing Practice for Medical Devices: General*, elaborated by the Food and Drug Administration (FDA). Published in 1978, this document is presently under revision, and this revision will again be consistent with the ISO 9000 family of standards. It also intends to be consistent with the International Standard ISO 13485 'Quality systems—Medical devices—Particular requirements for the application of ISO 9001'. As ISO 13485 follows a similar philosophy to that adopted in Europe through EN 46001, important steps towards international harmonization of quality standards for the field of medical devices and surgical products are now being taken—an essential benefit for an industry addressing highly specialized markets of worldwide extension.

AIR CLEANLINESS REQUIREMENTS

The air cleanliness requirements for medical device manufacturing facilities must inevitably reflect the extraordinary range of variation covered by this product family:

products vastly different in purpose, in size and shape, in the materials employed, in the manufacturing methods utilized, and in their exposure times to the production room atmosphere. Some are simple and manufactured by fully automatic procedures, such as syringes and wound dressings. Others are extremely complex assemblies, such as the microinvasive tools used in some procedures of modern surgery, or the electronic heart pacemakers. Others still employ extremely stubborn materials and have to be packed manually, such as the protein wound threads that dissolve in the body after they have rendered their service. For this reason, standards and guidelines refrain from being too specific in the determination of air cleanliness levels to be maintained in the manufacturing and assembly facilities. For example, the European Standard EN 46101:1993 simply states: 'The supplier shall establish and document requirements for the environment to which product is exposed'. Even the GMP guidelines do not proceed much beyond that. The *Blue Guide*, for example, states: 'This guidance does not attempt to specify the class of cleanroom which should be used. The cleanliness of the air should be defined for each type of area and will depend on the nature of the work, the product and the degree of handling and exposure of the product. Within these areas it may be desirable to use locally contained work stations or benches which provide cleaner air'. The American determinations in 21 CFR part 820 are more general still.

On the basis of such philosophies, the appropriate air cleanliness level is determined as result of a risk analysis. This takes into account factors such as the exposure time of the product and its components to the room atmosphere during processing, the amount and character of human interference with the product during manufacturing, assembly and testing, plus the efficiency of the final cleaning procedures employed after assembly. The inspection authorities, in turn, will assess both the plausibility of the chain of arguments behind the determinations, as well as the quality of the proofs presented to corroborate the claims.

For the manufacture and assembly of devices and surgical products permitting terminal sterilization, an air cleanliness level corresponding with Class 10 000 or 100 000 (ISO Class 7 or 8) will, as a rule be sufficient, as a certain limited bioburden can be tolerated prior to sterilization.

However, in the case of products which do not permit terminal sterilization after manufacture and assembly, extremely careful precautions have to be taken in order to avoid contamination of the products with viable microorganisms. In circumstances such as these, Class 100 (ISO Class 5) is normally maintained in the critical manufacturing zone, and Class 10 000 (ISO Class 7) in the controlled area surrounding it. These conditions largely correspond with those observed in the pharmaceutical industry during aseptic processing. There, of course, strict separation of the process from human interference is the rule. In medical device manufacturing, on the contrary, many of the delicate assembly and packaging operations of complex, obstinate device components require the manual skills of well-trained operators which cannot—at least not yet—be substituted by assembly or handling robots. The need for extreme motivation, discipline and responsibility required from the medical device assembly staff in such cases is indeed a challenge which has to be taken into account in this industry's quality systems.

CLEANROOM CONFIGURATIONS

Owing to the wide range of products and production procedures, no preferred clean-room designs can be identified for the medical device field. Concepts inevitably vary widely, as they are determined individually according to each specific situation. There-fore, typical examples will illustrate some of the many options utilized for meeting the given contamination control objectives not only effectively, but also economically (Schicht, 1994).

Injection Cannulae

The manufacture of injection cannulae is an example of a fully automated production process employing machinery which has neither been developed specifically for medi-cal device manufacturing nor for operation in a clean environment (Frei, 1993). The injection moulding machines for producing the cannula components are arranged as clean islands maintained at Class 1000 (ISO Class 6), within a hall maintained at Class 100 000 (ISO Class 8) air cleanliness. Figure 5.1 shows one of these islands which is shielded against its surroundings by means of plastic curtains. The supply air outlets are positioned directly above the injection moulding machines. The curtains end 10 cm above floor level, permitting the clean air to spill over into the surroundings. The Class 100 000 (ISO Class 8) air cleanliness maintained there is thus obtained, as it were, as a side-effect and at no additional cost. Assembly of the cannula components into the final product is again fully automatic and it performed in an adjacent Class 10 000 (ISO Class 7) cleanroom of similar concept.

As a consequence of the strict elimination of human interference during the entire production sequence, the bioburden of the final product prior to terminal sterilization is extremely low.

Heart Pacemakers

The heart pacemaker is an excellent example of a highly sophisticated micro-electronic device which has to comply with particularly severe reliability and durability criteria. The electronic circuits must therefore be protected systematically against the risk of malfunctions which may be caused by particulate contamination or by corrosion effects. Therefore, component manufacture and the entire assembly operation must be performed under cleanroom conditions satisfying high air cleanliness levels and employing manufacturing procedures with minimum particle generation. For products of this type, particles of any kind—not just microorganisms—have to be controlled. As special precautions against corrosion risks, vacuum drying procedures for the micro-electronic circuitry, and fully automated final assembly in isolators in which a nitrogen atmosphere with less than 1% relative humidity is maintained, were employed in the example described (Perrenoud, 1993). After assembly, the pacemaker units are her-metically sealed by means of laser welding in a helium atmosphere.

Figure 5.2 shows the assembly and computer-assisted remote electrical testing of pacemaker electronic modules. The isolator concept of the sealed assembly lines and the glove ports for manual process interference are clearly discernible. An air cleanli-ness level according to Class 100 (ISO Class 5) is maintained in the assembly isolators

FIGURE 5.1. Cleanroom with turbulent airflow for manufacturing injection cannula components by means of an injection moulding procedure (courtesy: Disetronic Ltd, Burgdorf, Switzerland).

FIGURE 5.2. Assembly of heart pacemakers in an inert gas isolator with its adjacent preparation and inspection area. The air cleanliness corresponding to Class 100 (ISO Class 5) is maintained by means of vertical displacement airflow (courtesy: Sulzer Intermedics SA, Le Locle, Switzerland).

FIGURE 5.3. Cleanroom equipped with horizontal displacement airflow clean workstations for the preparation of heart valve grafts from porcine aortic tissue (courtesy: Baxter Edwards Ltd, Horw, Switzerland.

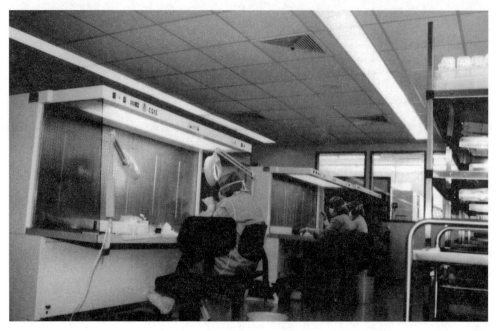

FIGURE 5.4. Manual sewing operations are an essential feature of the preparation of heart valve grafts from animal tissue (courtesy: Baxter Edwards Ltd, Horw, Switzerland).

and at the inspection work stations, with Class 1000 (ISO Class 6) conditions applying to the circulation areas. All the usual additional precautions such as cleanroom garments for the personnel as well as the use of air locks for the transfer of persons and material into the critical area are also employed.

Aortic Bioprostheses

The preparation of heart valve grafts from animal tissue is an appropriate example of medical devices requiring aseptic handling. As the products cannot be submitted to terminal sterilization, ample dispositions must be taken to avoid microbial contamination during the entire processing sequence. Figure 5.3 shows a cleanroom employed for this task. The workstations, where the tissue material is converted into the finished product by means of manual sewing operations, have to obey Class 100 (ISO Class 5) air cleanliness conditions, whereas the background is maintained at Class 10 000 (ISO Class 7). In such a work procedure relying on manual labour, particular attention must be paid to microbial dissemination from the employees' bodies which even with the best cleanroom garment concepts cannot be entirely suppressed. Therefore, as Figure 5.4 shows, clean workstations with horizontal airflow are employed. The HEPA filtered air is directed horizontally against the operators and all microorganisms disseminated by them are swept away from the critical manufacturing area into the circulation area behind their backs where they can cause no harm. Frequent decontamination of the work area and monitoring of the particle count in the *operational* mode of the facility are important contributions to quality assurance in such a delicate manufacturing operation.

CONCLUDING REMARKS

The case studies presented illustrate the individual approach appropriate for determining the conceptual design of cleanrooms for medical device and surgical product manufacturing. Each case was distinguished by individual air cleanliness determination, and by individual assessment of the specific risk situation. But they have one feature in common: consequent limitation of the air cleanliness controlled zones to the critical process areas. Thus, the pre-determined performance criteria are met efficiently and economically.

REFERENCES

International Standard ISO 13485. 'Quality systems—Medical devices—Particular requirements for the application of ISO 9001'. International Organization for Standardization, Geneva (Dec. 1996).

European Standard EN 46001 (1993). 'Quality systems—Medical devices—Particular requirements for the application of EN ISO 9001'. European Committee for Standardization, Bruxelles.

Federal Regulation 21 CFR 820. *Good Manufacturing Practice for Medical Devices: General.* U.S. Government Printing Office, Washington.

Frei, Th. (1993). 'Zum Einsatz von Kunststoffspritzgiessmaschinen und Montageautomaten im Reinraum bei der Herstellung von Injektionskanülen (The application of injection-moulding machinery and automated assembly in cleanrooms during injection cannula production)', *Swiss Med.*, **8**-S, 21–24.

Perrenoud, J. J. (1993). 'Assurance qualité et technique des salles propres dans la pratique industrielle—Les stimulateurs cardiaques comme cas pratique de la technique médicale (Quality assurance and cleanroom technology in industrial practice—Heart pacemakers as example from medical device technology)', *Swiss Med.*, **8**-S, 25–26.

Schicht, H. H. (1994). 'The use of clean-room technology in medical device manufacturing: A tool in the service of quality', *Medical Device Technology*, **5**, 2, 22–26 (part I); **5**, 4, 18–22 (part II).

6 Contamination Control Facilities for the Biotechnology Industry

P. J. TUBITO and T. J. LATHAM

INTRODUCTION

Modern biotechnology developed in the late 1970s, arising from advances made in molecular and cellular biology and their application to industrial processes. It built upon techniques developed in traditional biotechnology industries, such as antibiotic and vaccine manufacture, that exploited the properties of naturally occurring micro-organisms. Modern biotechnology could 'create' new organisms utilizing the techniques of genetic engineering. The potential escape of these new recombinant organisms from the laboratory to the outside environment was a concern both to scientists and to public administrations. From the very start, biocontainment was an issue that had to be addressed in the design and operation of modern biotechnology facilities.

Traditional biotechnology is used in a wide range of industries, including the manufacture of food and food additives, brewing, the production of bulk chemicals such as acetone, and the manufacture of pharmaceuticals, including antibiotics, steroids, vitamins, vaccines and others. Initially, it was expected that genetic engineering could be applied broadly across the range of these industries, but in the last 20 years the greatest and widest range of applications has been found in pharmaceutical manufacture. Since most 'biopharmaceuticals' are intended for parenteral administration and thus need to be sterile, this has led to the need to design manufacturing facilities that incorporate the dual requirements of biocontainment and asepsis. Such facilities must prevent the contamination of the outside environment by the recombinant organism, whilst also preventing contamination of the product itself by organisms or particulates. Often conflicting practical and regulatory requirements must be resolved to achieve a workable facility.

The aim of this chapter is to describe the contamination control facilities used in biopharmaceutical manufacturing plants. To do this, it is first necessary to describe the nature and typical features of bioprocess operations. Since the general influence of Good Manufacturing Practices (GMP) on pharmaceutical cleanrooms has already been described in Chapter 4, we concentrate largely on biocontainment and the integration of this with GMP requirements.

Cleanroom Design. Edited by W. Whyte © 1999 John Wiley & Sons Ltd

BIOTECHNOLOGY—THE INDUSTRY

It was only in 1973 that foreign DNA fragments were functionally incorporated into a cell structure. This breakthrough demonstrated that the selective, deliberate alteration of the hereditary code of a living cell was possible. Recombinant DNA makes it possible to modify cells to produce a desired product. As an example, rDNA techniques have been used to modify the genetic makeup of a bacterium to enable it to produce human growth hormone. Prior to this innovation, this product could only be obtained in limited quantities from the pituitary glands of human cadavers. The product could potentially be contaminated by human viruses that had infected the source—Creutzfeldt–Jacob virus being a particular concern. This illustrates one advantage of biopharmaceutical products: their freedom from contamination by viruses such as hepatitis and HIV. Another advantage is the ability to manufacture large quantities of human proteins unobtainable by other means.

Monoclonal antibody technology, which uses a fused product of an antibody-producing cell and a cancerous cell called a hybridoma, was developed in 1975. The hybridoma cell can be designed to produce large quantities of a specific antibody, and this technology is now exploited for the manufacture of diagnostics and therapeutic proteins.

The first products derived from these new technologies and approved by the regulatory agencies were human insulin (1982) and human growth hormone (1985). These were closely followed by an interferon (1986), hepatitis B vaccine (1986), tissue plasminogen activator (1987), interleukin 2 (1988) and erthropoietin (1989). All of these pharmaceutical products were supplied for use as injectables. In recent years, further products have been licensed. Currently, over 150 recombinant protein products have been developed to the stage of clinical trials or beyond.

BIOPROCESS OPERATIONS

Prior to discussing biocontainment it is necessary to describe the nature of the processing operations used in biopharmaceutical manufacture. These will, of course, vary according to the nature of the product and the producing organism, the scale of manufacture and the ultimate function of the product and its dosage form. However, there are typical methods and operations used across the broad spectrum of biopharmaceutical manufacturing that are different from those used for chemically synthesized pharmaceuticals.

Typically, a biopharmaceutical manufacturing process can be divided into four major steps: media preparation, fermentation, product recovery and purification, and finishing. These are illustrated in Figure 6.1.

Media Preparation

Media preparation involves the dispensing of raw materials from bulk supplies, which may be liquid or solid, the formulation of defined media solutions and their subsequent sterilization. When the organism is a yeast, fungus or bacterium, the media components will mostly be solids. Usually, they will include a carbohydrate such as dextrose or sucrose, a nitrogen source such as yeast extract and smaller quantities of materials

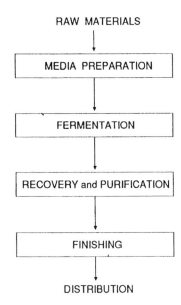

FIGURE 6.1. Bioprocessing steps.

supplying vitamins, minerals and essential growth factors. Processes utilizing animal cells or hybridomas have more complex nutritional requirements, the media being essentially a protein solution that is often based on bovine serum albumin. Such media may be supplied in either liquid or powdered form.

The nature of the media and its components defines the requirements of media preparation areas. These will usually include means of weighing solids, metering liquids, mixing and dissolving these components. The area will be supplied with some form of purified water. The formulated media will be sterilized, nowadays most commonly by filtration but possibly by heat in the case of solids-containing fermentation media.

The presence of dusts from solids-batching operations means that some form of environmental control facility is required to protect operators and prevent cross-contamination. Conventional dust extract systems may be used, or fume cupboards. On a large scale, unidirectional downflow booths are quite popular. Where biopharmaceutical GMP standards apply, the operation will be carried out in a suitable environment, usually class 10 000, to prevent contamination of the media by extraneous material.

The culture organism has not yet been introduced into the process, so biocontainment is not an issue for media preparation areas.

Fermentation

Fermentation is the process by which a micro-organism grows and converts raw materials into products. The micro-organisms may be a yeast, fungus or bacterium. Where an animal cell or hybridoma is used as the production organism, the process is usually referred to as cell culture.

Fermentation requires that the micro-organism or cell be provided with an environment suitable for optimal growth and product formation. This is provided in the form of a bioreactor or fermenter. This is a closed vessel that can be sterilized and into which the sterile media is injected. The media is then aseptically inoculated with the production organism. The bioreactor allows control of the liquid environment in terms of temperature, pH, oxygen content and other parameters. It also allows aseptic operation such that a pure culture of the production organism is maintained.

There are various designs of bioreactor, but mostly they are agitated jacketed vessels fabricated from stainless steel. For biopharmaceutical manufacture they typically range from 1000 to 50 000 litres. They have sophisticated instrumentation and control systems. A typical example is shown in Figure 6.2. The maintenance of asepsis and the containment of the hazardous organisms requires special attention due to the large number of possible penetration points. These are primarily the exhaust air vent, the agitator seal and static seals at valve and pipe or flange joints.

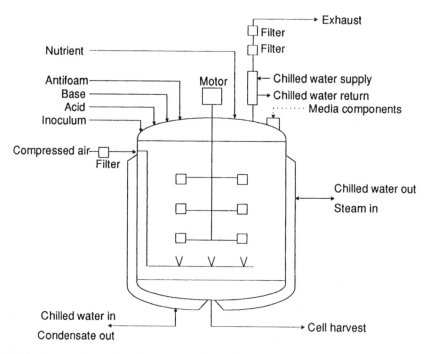

FIGURE 6.2. Suspension culture fermenter schematic.

Recovery and Purification

In many bioprocessing operations the living micro-organism has fulfilled its useful purpose by the end of fermentation, and so it is killed. This may be achieved by chemical or thermal treatment in the bioreactor, or by filtration of the culture fluid as it is discharged from the bioreactor. In such cases, biocontainment may not be required for subsequent processing operations.

It is always preferred that the micro-organism is contained within the bioreactor.

TABLE 6.1. Unit operations for the recovery and purification of proteins.

Function	Unit operation
Cell harvesting	Disk-stack centrifuges, cross-flow microfilters
Lysis	homogenizers, ball mills
Debris removal	Disk-stack centrifuges, cross-flow microfilters
Crude purification	Precipitation (with salts, solvent, pH shift)
High resolution	Chromatography
Product concentration	Ultrafilters (UF)
Desalting	UF, gel filtration chromatography
Precipitate recovery	Centrifuges (tubular bowl, disc-stack)
Sterilization	0.22 μm filters

There are many instances, however, where biocontainment must be extended to include recovery and purification processes. This may be because the cells cannot be killed because of potential damage to the product or disruption to the efficiency of the processing operation. It may also occur if even the killed cells have a potentially toxic effect. This situation often arises in vaccine manufacture, where antigenic toxins are being produced. Even relatively innocuous micro-organisms, such as those used for producing enzymes for biological washing powders, have been found to cause a potentially harmful immune reaction in some process operators.

The typical unit operations of protein recovery are described in Table 6.1. The initial steps are dependent upon whether the product is intracellular (produced and retained within the cell) or extracellular (excreted into the fermentation broth). If the recombinant organism has not been killed in the bioreactor, then the initial recovery steps may have to be performed under biocontainment.

For large-scale recombinant fermentations producing an intracellular product, the initial recovery of the cells from the fermentation broth is typically performed in a disk-stack centrifuge. The recovered cells are then broken apart (lysed) in an homogenizer or bead mill to release the product. The remaining unit operations are selected according to the properties of the product and the contaminating molecules, which are either left over from the fermentation medium or released from the cells during lysis. The aim of the initial recovery processing is the high efficiency recovery of the product protein with maximum rejection of unwanted material, preferably also reducing the volume of the process fluid.

A complete discussion of the various unit operations is beyond the scope of this chapter. It will, however, be useful to describe those operations in which biocontainment is frequently needed, and these are largely the initial recovery steps after the fermenter which may still contain live or dead cells, or substantial quantities of impurities.

Centrifugation   Disk-stack centrifuges are most frequently used for separating cells from their fermentation medium. Usually machines are used that can continuously discharge the cell paste, and thus centrifugation of the contents of any particular fermenter is a continuous process. Machines are available for biocontainment operations, being provided with 0.22 μm filters to prevent aerosol release from vent pipes. They have facilities for steam sterilization to enable the centrifuge to be decontaminated at the end of the operation.

Tubular bowl centrifuges are sometimes used for cell separation, but are more commonly used to recover protein inclusion bodies (solid protein aggregates formed within bacterial cells). Biocontainment is difficult because these machines release aerosols from their shaft seals during operation. Additionally, they retain the cells in the bowl, which has to be dismantled at the end of the operation to recover the product. They are therefore not recommended for biocontainment operations. If necessary, they may be housed in flexible film isolators to contain aerosol release, and their dismantling may be done by operators wearing breathing suits.

Homogenization   Homogenizers break cells apart by pumping the broth at high pressure across a letdown valve. The broth may be circulated several times through the homogenizer, each pass releasing increasing quantities of protein. The broth must be cooled to remove the heat generated by the pump. Studies have shown that homogenizers release aerosols containing cell fragments, so for biocontainment the whole machine must be enclosed in an isolator. Methods have been developed for decontaminating the homogenizer by steam after use. The internal surfaces of the isolator must also be decontaminated prior to personnel entry, and this may be achieved by some form of vapourized disinfectant, either formaldehyde, peracetic acid or hydrogen peroxide.

Microfiltration   Cell recovery can be accomplished using microfiltration systems, which employ membranes made from a variety of materials including ceramics and fluoropolymers. These membranes can be in spiral, flat or tubular cartridges. Filtration is achieved at relatively low pressures, a cell concentrate being retained by the membrane separated from a cell-free permeate. Only a few of the available membranes are steam-sterilizable, but chemical disinfection can be used for decontaminating the non-steamable membranes. The typical disinfectants used are peracetic acid, hydrogen peroxide and caustic solutions.

Finishing

The final formulation and packaging is performed in the finishing area. Sterile products are processed in a similar manner to conventional sterile pharmaceuticals (see Chapter 4). Some of the typical operations performed in the finishing area are described in Table 6.2.

TABLE 6.2. Unit operations for the finishing of proteins.

Function	Unit operation
Solution preparation	Stoppering/capping
Sterile filtration	Lyophilization (freeze drying)
Terminal sterilization	Inspection
Component preparation	Labelling/coding
Aseptic filling	Packaging

At this stage of processing there is rarely a requirement for biocontainment. However, some form of containment is usually necessary to prevent the exposure of operators to pharmacologically active materials. The need to protect the product from contamination usually requires features that also provide the necessary level of containment, and these will have a significant impact on the design of the facility, particularly in the usual case where the product is finished in a sterile dosage form.

The regulatory agencies prefer that aseptic filling be carried out in a dedicated building separate from that which houses a biocontainment area. If this is not feasible due to cost or space limitations, then the filling facility can be isolated from the biocontainment area in the same building. Thus, they are provided with separate access routes, separate air handling systems and dedicated equipment, utility services and glassware.

Utility Services

Utility services that may potentially contact the process fluid containing the biopharmaceutical product must be of such a quality that contamination is prevented or minimized. Therefore, those services that enter the processing equipment, such as process air, sterilizing steam and process water, are supplied in purified form.

It is not permissible under GMP standards to use non-purified water in biopharmaceutical manufacturing. At the very least, water must be under some form of chemical and microbiological control, and usually a pharmacopoeial grade of water is required. US and European Pharmacopoeias define standards for Purified Water (PW) and Water for Injection (WFI). The latter is the higher standard and is used in the later stages of the processing of parenteral products. PW is produced usually by a series of purification methods, which may include carbon filtration, softening, deionization and membrane processes. WFI is usually produced by distillation, although reverse osmosis is also used in the United States and ultrafiltration is permissible in Japan. PW may be distributed through stainless steel or plastic pipe systems, designed to be sanitizable. WFI is always distributed through stainless steel pipe systems that are designed to be steam-sterilizable. In both cases, the pipe systems are designed as a recirculating loop to reduce microbial proliferation should contamination occur. Often in the case of WFI, the loop is run hot (60–80°C) to decrease further the opportunity for microbial growth.

Steam is very often injected into biopharmaceutical processing equipment as part of the sterilization process, and any contaminant in the steam could remain in the equipment and contaminate the pharmaceutical product. Therefore, contaminant-free steam must be supplied, and is commonly known as clean steam. Conventional plant steam, used for HVAC heating, has additives to prevent boiler scaling and corrosion. These cannot be added to the water used for raising clean steam, because they would be carried over into the steam. The feedwater is usually treated to remove dissolved chemicals that cause scaling and corrosion, and special stainless steel steam generators are used to provide further corrosion resistance. The feedwater may be softened, deionized or purified water, of good microbiological quality. The clean steam quality is usually defined such that the condensate is equivalent to the pharmacopoeial grade of water used in the facility. Clean steam that produces condensate equal to WFI quality is often referred to as pure steam.

Air, and sometimes other gases (nitrogen, CO_2, oxygen), are sometimes injected into process equipment, to feed growing micro-organisms, to effect pressure transfers or to blow dry equipment after cleaning. The gas is usually sterilized by filtering through 0.22 μm filters prior to entering the facility, and is then distributed through steam-sterilizable stainless steel piping to the user points. If the air is supplied from a compressor rather than cylinders, precautions may be required to prevent contamination from the compressor. An oil-free compressor may be used, but this does not deal with hydrocarbon contaminants from the atmosphere. Perhaps the better solution is to use a conventional oil-lubricated compressor with adequate downstream purification filters. This is a more reliable, more energy efficient and less costly system than oil-free systems, and can be designed to deal adequately with atmospheric contaminants.

There may be certain instances where other services must be treated to prevent contamination of the cleanroom environment. For instance, lubricated air used for powering air diaphragm pumps must be piped outside the cleanroom, otherwise an oil film can quickly arise in the area around the pump.

BIOCONTAINMENT

Biocontainment was an issue of concern for microbiologists for many years preceding the development of recombinant DNA techniques. It was required for use with potentially hazardous natural organisms that might be encountered in general research, medical diagnosis and the manufacturing of vaccines, amongst other things. The advent of genetic engineering was accompanied by a fear that new 'unnatural' micro-organisms might be released into the environment, with unpredictable effect. Biocontainment of recombinant micro-organisms was addressed very early in their development, and the first guidelines were published by the National Institutes of Health in the United States in 1976. These have subsequently been revised, and to them has been added a range of international and national guidelines.

Biocontainment Legislation

In the United States   US requirements for biocontainment of Genetically Modified Organisms (GMOs) are given in the revised NIH Guidelines, published in the Federal Register of 18 July 1991 (56 FR 33174) and in force since then. These guidelines offer four standards of laboratory containment, Biosafety Levels 1 to 4 (BL1 to 4). For experiments at greater than 10 litres scale, four additional biocontainment categories are offered. For non-pathogenic, non-toxigenic recombinant organisms, or those that are unable to survive outside controlled laboratory conditions, the Good Industrial Large Scale Practice (GLSP) level is recommended. Above this are Biosafety Levels— Large-Scale (BL1-LS, BL2-LS and BL3-LS) each the equivalent to BL1 to 3. There is no standard provision for the large-scale culture of the very hazardous organisms that require BL4 biocontainment.

Table 6.3 shows the equivalent biocontainment standards for the US, European and UK legislation. The specific requirements of the US regulations are given in Table 6.4.

In Europe   In the countries of the European Union, biocontainment of GMOs is required to meet the standards of the EC Directive 90/219/EEC of 1990. In the United Kingdom, these have been formulated into the Genetically Modified Organisms

Table 6.3. Equivalent biocontainment standards for large-scale operations.

	US NIH	EC Directive	UKGMO
No biohazard	GLSP		
Low biohazard	BL1-LS	LSCC-1	B2
Medium biohazard	BL2-LS	LSCC-2	B3
High biohazard	BL3-LS	LSCC-3	B4

Table 6.4. Facility design requirements of US biocontainment measures for large-scale operations with potentially hazardous genetically manipulated micro-organisms.

Specifications	NIH containment levels			
	GLSP	BL1-LS	BL2-LS	BL3-LS
1. Airlocks at suite entrances				Required
2. Biohazard signs at suite entrances			Required	Required
3. Negative room air pressure				Required
4. Room finishes to suit easy surface decontamination				Required Required
5. Decontamination facilities for personnel				Required
6. HEPA filtration of exhaust air from the suite				
7. Closed systems for all operations with viable organisms		Required	Required	Required
8. Treatment of exhaust process gases from closed systems		Minimize release	Prevent release	Prevent release
9. Closed system to be operated at as low a pressure as possible				Required
10. Containment of rotating seals and other penetrations of the closed system			Prevent release	Prevent Release
11. Instrument monitoring of the integrity of the closed system			Required	Required
12. Validated integrity testing of the closed system			Required	Required
13. Validated inactivation of organisms prior to removal from the closed system		Required	Required	Required
14. Control of aerosols during process operations	Minimize release with procedural controls	Minimize release with engineering controls	Prevent release	Prevent release
15. Inactivation of waste materials		Required	Required	Required
16. Emergency systems/procedures for handling accidental spillages	Recommended	Required	Required	Required

(Contained Use) Regulations, issued as a Statutory Instrument by the Health and Safety Executive and in force since February 1993. This offers methods for classifying organisms according to their potential hazard, and three levels of biocontainment (B2, B3 and B4) for large-scale production of those recombinant organisms that are pathogenic or toxigenic and are capable of surviving outside the controlled environment. These are identical to Large-Scale Containment Categories (LSCC) 1, 2 and 3 of the EC Directive.

UK requirements for the containment of naturally occuring micro-organisms are given in *Categorization of Pathogens According to Hazard and Categories of Containment*, 4th edn, 1995, published by the Advisory Committee on Dangerous Pathogens (ACDP). This categorizes micro-organisms according to four levels of potential hazard, from Group 1 (unlikely to cause human disease) to Group 4 (capable of causing an untreatable, frequently fatal disease, which may spread through the community). Four containment levels are defined, CL 1 to 4, each appropriate for the equivalent hazard group. They are aimed primarily at laboratories and animal houses rather than process equipment. They are therefore of less use than GMO regulations for the design of biopharmaceutical manufacturing facilities.

Recently, a British Standard/EuroNorm has been issued for biotechnology containment facilities, and provides a common standard required to be met in all the member states of the European Community. This is BS/EN 1620 (1997), 'Biotechnology—Large-scale process and production—Plant building according to the degree of hazard'. This lists four containment categories similar to those of the ACDP listed above. Additionally, it provides for classification of micro-organisms according to their potential hazard to humans, animals or plants.

Primary Containment

Items 1–5 in Table 6.4 are measures of primary containment in that they aim to contain the hazard within a 'closed system' of processing equipment. It is impossible to create a completely closed system due to the need to supply feedstocks, remove samples, fill and vent the system, etc. Therefore, provisions must be made to minimize or prevent the release of micro-organisms during these operations.

The following is a list of features that should be considered for incorporation into the equipment design of a contained system:

- All vessels and equipment containing live organisms should be suitable for steam sterilization. Preferably, they should be designed for sterilization-in-place (SIP) rather than relying on dismantling for sterilization in an autoclave.
- Exhaust gases from contained processing vessels such as bioreactors must be passed through 0.2 μm sterilizing filters. Under certain circumstances, the exhaust air may also require incineration.
- Special consideration must be given to the design of seals on penetrations into the closed system. Flange joints, instrument probes and other static joints usually require a single O-ring at the lower levels of containment. Vessels used for B4 containment may require double O-rings or double O-rings with steam flushed in the annulus.
- Dynamic joints, such as agitator shaft seals, may require a double mechanical seal, flushed with a sterilizing agent such as steam or condensate.

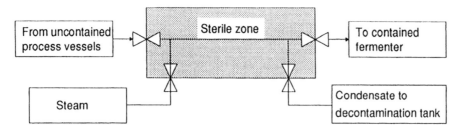

FIGURE 6.3. Steam barrier.

- During operation, a steam barrier can be maintained on fixed piping leading to the contained vessel. Figure 6.3 illustrates the operation of a steam barrier. During a transfer, the steam and condensate valves are closed and the process fluid can be transferred from the uncontained vessel to the fermenter. During fermentation, the two valves in the process transfer line are closed and steam is introduced in the zone between the two valves. This steam barrier is maintained throughout the fermentation to ensure that the bioreactor will not be contaminated and that the contained organism will not escape.

- Potentially contaminated waste streams such as sampling waste and steam condensate from equipment sterilization must be routed to a decontamination tank for sterilization prior to discharge. Contaminated process waste streams such as centrifuge supernatant must also be routed to the decontamination tank.

In cases where the equipment cannot be designed as a closed system, such as when removing cell paste from a tubular bowl centrifuge, the room itself may be regarded as the primary containment. In such cases, the operators within the room must be protected by breathing suits.

Secondary Containment

The primary containment of a bioprocessing operation may be breached by various means, such as leakage, equipment failure and maloperation. It is necessary, therefore, to provide secondary containment, which is largely facility-orientated. It is designed to prevent any organisms released from the contained process plant from passing outside the facility to the outside environment.

The major implication of secondary containment is that the facility actually becomes part of the process operation. A great deal of co-ordination is therefore needed between the various engineering disciplines during the design of a facility to ensure that the overall containment concepts are applied consistently to both the process equipment and the building.

The design implications of secondary containment to the architectural discipline range from room layout to the selection of room finishes. The following identifies some of the established design practices:

- It is desirable to separate physically the biocontainment area from other building functions. The entry should not be from an unrestricted corridor. Containment categories BL1-LS and BL2-LS may require an area for gowning and washing.

BL3-LS containment requires a change area with a personnel shower that provides access via an airlock.

- The biocontainment area must also be designed to contain an accidental spillage. This may be accomplished by having the floor recessed slightly below adjacent uncontained areas. The floor should be seamless, coved at the borders and sloped to the drains. Materials that are typically used include welded sheet vinyl, trowelled-on epoxy or terrazzo.

- The floor drains should be routed to the waste collection vessel for decontamination.

- The materials selected for walls and ceilings must be water resistant so they can be easily cleaned. Walls must also be chip resistant if mobile tanks are to be wheeled around. Typical wall finishes include gypsum board with either an epoxy paint or PVC coating and an epoxy paint over a block wall with a plaster filler.

- BL3-LS containment areas must be completely sealable to allow fumigation by a vapour-phase sterilant such as formaldehyde.

Some of the facility design concepts described above are demonstrated in the room layout presented as Figure 6.4.

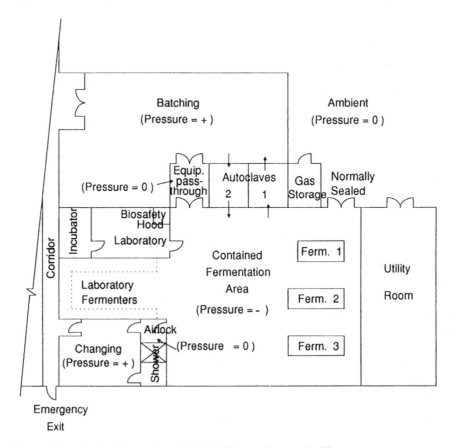

FIGURE 6.4. Hypothetical layout for a BL3-LS biocontainment facility.

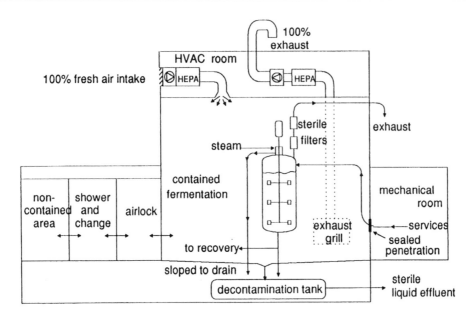

FIGURE 6.5. Section through a BL3-LS biocontainment facility.

The design implications of secondary containment that must be considered with respect to building services are the two methods used to isolate the environment within the secondary containment area from the outside environment. These are High Efficiency Particular Air (HEPA) filters and room pressure differentials. HEPA filters are required for air being exhausted from the higher containment categories. Supply air to the room may also be HEPA filtered in order to decrease generally the bio-burden within the room and increase the life of the exhaust air HEPA filters. Those HEPA filters that can be potentially contaminated should be installed in 'safe-change' housings so that the person changing the filter will not be exposed to the organisms.

BL3-LS containment requires that the area be maintained at a negative pressure relative to the adjacent area. Airlocks located between those areas will be maintained at approximately one-half of the pressure differential. This encourages airflow from the uncontained area to the contained area. The design concepts of primary and secondary containment as they apply to a typical BL3-LS fermentation area are illustrated in the building section shown in Figure 6.5.

Decontamination of Liquid Wastes

All the NIH large-scale biocontainment levels require that liquid waste be inactivated prior to release from the facility (Table 6.4, items 13 and 15). This is generally accomplished by collecting the waste in a contained vessel, then treating it either with a disinfectant or by heat to kill all recombinant micro-organisms present. The collected waste will include not only the process effluent, but also any liquid that may have been contaminated by the micro-organisms within the facility. This could include seal water, condensate from sterilization operations, wash water and other fluids.

Two main forms of waste inactivation are used. Batch inactivation requires a mini-

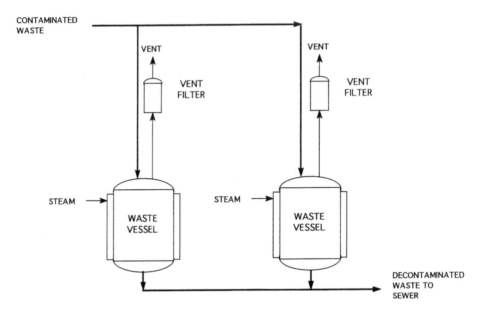

FIGURE 6.6. Batch decontamination of biohazardous waste.

mum of two waste collection vessels (Figure 6.6). These are filled alternately, and as each becomes full it is closed off and the waste collection function is transferred to the other vessel. The full vessel is then subjected to the inactivation procedure. Chemical disinfection requires that a chemical be added to the vessel, and then a relatively long holding time is required for this to take effect. Some micro-organisms are inactivated by high pH, so sodium hydroxide solution can be added to the vessel. For more resilient micro-organisms, proprietary disinfectants may be used.

Thermal disinfection is perhaps more common, and requires that the vessel contents be heated and held at a high temperature. Micro-organisms differ in their ability to withstand high temperatures and an appropriate temperature and holding time must be established by prior experimentation. Usually, temperatures between $80°$ and $100°C$ are used, with holding times of several minutes to one hour.

Continuous inactivation is a second form of waste inactivation. It usually involves a thermal method of inactivation, and requires only a single waste collection vessel (Figure 6.7). As this vessel fills, its contents are pumped through a heater and then through a holding coil or vessel, this system being designed to bring the waste to the appropriate temperature and hold it there for the required time. Usually, the temperatures used are higher than those of batch disinfection, but the holding times are consequently shorter.

BIOPHARMACEUTICAL MANUFACTURING FACILITIES

The biopharmaceutical manufacturing facility will include the main processing areas described previously. Many modern facilities are designed for multiproduct operation, such that there will be a separate suite of manufacturing areas for each product. There

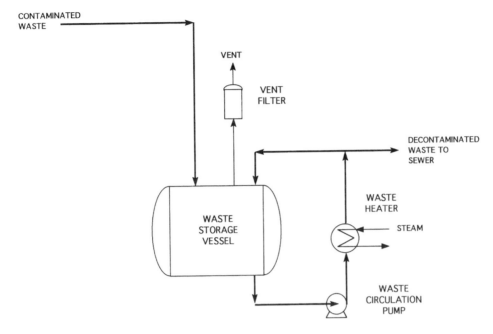

FIGURE 6.7. Continuous decontamination of biohazardous waste.

will be access and communication routes within the facility, together with the support areas required. These may include:

- Laboratories for process analysis, quality control, inoculum development.
- Utility and mechanical service rooms, service corridors for the distribution of piped services and interstitial floors for ducted services.
- Stores for equipment and materials. These may include cold rooms.
- Warehouses for raw materials and product.
- Locker rooms, changing rooms and toilets.
- Administrative facilities such as offices, archive stores, conference rooms.

The facility must be designed specifically to protect the worker, product and environment. Particular features that must be incorporated into the facility design include the integration of biocontainment and product isolation, security from fire and explosion hazards and precautions against operator exposure to radioactivity.

Integration of Biocontainment and Product Isolation

The GMP requirements relating to cleanrooms are described in Chapter 4. One of the key requirements is the use of positive pressure differentials in cleanrooms to prevent contamination from the external environment. This directly contradicts the biocontainment requirement for negative pressure differentials. Means have to be found to satisfy both requirements. Generally, the requirement for the area around the process equipment to be at negative pressure is overriding. This need not necessarily be the process room itself, but can be a flexible film isolator, separately ventilated. Where equipment is too large or complex for such a solution, then the biocontainment suite (at negative pressure) must be contained within a surrounding building envelope at positive pressure. Air flow within the facility must be designed such that it is exhausted at an intermediate zone, usually an airlock, between the two areas.

Fire and Explosion

The design of biopharmaceutical facilities must include the normal building regulation requirements for limiting fire spread and assisting personnel escape. Some facilities must incorporate additional specialized requirements for flammable operations. Such operations are normally associated with the use of flammable solvents. These are not widely used in bioprocessing, but ethanol, isopropanol, acetone and other solvents are sometimes used as precipitating agents, for selective dissolution and other activities during recovery and purification operations.

The following are some of the design considerations:

- A segregated storage and dispensing area should be provided, incorporating drainage systems to contain any liquid spillage.
- Structures should be fire resistant.
- Mechanical exhaust ventilation should be provided, complete with controls and safeguards, to prevent build-up of flammable atmospheres.
- A high conductivity (anti-static) non-slip solvent-resistant floor should be installed.
- It may be a requirement under some national guidelines (e.g. NFPA in the United States) and some insurance company codes (e.g. Factory Mutual) to provide explosion relief in the building's walls or roof.
- Electrical installations are required.
- Process equipment may be provided with inert gas purging and blanketing.

Radioactivity

The production of some medical diagnostic products requires the use of radioactive tracers. The concentrated solutions of radio-isotopes must be handled in an area of the facility specifically designed for radioactive materials. The major design considerations include the following:

- Concentrated solutions of radio-isotopes should be handled within specially designed gloveboxes and hoods.
- Access to the room should be restricted with the entry via an airlock.
- The room must be maintained at negative pressure with respect to the adjacent areas.

- Mechanical ventilation of the room is required, utilizing once-through air (no recirculation).
- Exhaust air from the area should be filtered with activated carbon followed by HEPA filters to ensure that there is no radioactive emission to the outside environment.

CONCLUDING REMARKS

In the years since the publication of the first edition of this book, biocontainment has become a less emotive subject amongst scientists, engineers and the regulatory authorities at least. Most genetically engineered organisms are produced from non-hazardous natural organisms and agents. Additionally, they are either purposely weakened such that they cannot survive outside the controlled environment of the fermenter, or their metabolism is so overburdened by the changes made to it that they cannot compete with natural organisms. In either case, the hazard to the environment is reduced and this has been recognized by the regulatory authorities. Probably most biopharmaceutical facilities being built today are classified at the lowest level of containment. Nevertheless, stringent biosafety is still an issue in many facilities, and the use of recombinant bacteria and yeasts for vaccine manufacture, in particular, will ensure that it will continue to be so.

7 Cost-Efficiency and Energy-Saving Concepts for Cleanrooms

H. H. SCHICHT

INTRODUCTION

Cleanroom design is guided by the following priorities:

- first priority: its effectiveness in achieving the stipulated environmental conditions and required cleanliness class;
- second priority: functional safety and reliability;
- third priority: cost-efficiency, both as regards capital and operating cost.

This order of priorities is a consequence of the nature of the activities performed in the clean area, and of that fact that cleanrooms are normally equipped with extremely expensive production and testing apparatus, the correct operation of which must always remain guaranteed. It is furthermore a consequence of the fact that after a contamination problem has occurred, an extremely time-consuming requalification and revalidation of the process may be required.

Functional quality and operational safety are therefore the common denominators to which all suppliers tendering for a given cleanroom job have to comply. In the end, however, the commercial decision in favour of a certain supplier will always very strongly, if not decisively, be influenced by the aspects of investment and operating cost.

MINIMIZING THE AIR FLOW RATE FOR OPTIMUM COST-EFFICIENCY

Unidirectional air flow, which is also termed low-turbulence displacement flow in the European literature, and used to be termed—succinctly, but not quite correctly—laminar flow in the British and American technical literature, has shown in numerous and very varied fields of application that it is able to comply with the highest cleanliness requirements posed at present and in the foreseeable future. Unfortunately, this high performance does not come cheaply: with the commonly used air velocities of 0.3–0.5 m/s, an air flow rate of 1080–1800 m^3/h per square metre of filter area will result.

Cleanroom Design. Edited by W. Whyte © 1999 John Wiley & Sons Ltd

Unidirectional flow is therefore necessarily coupled with high volumetric air flow rates. As a consequence, the cost of systems employing unidirectional flow is intrinsically high, as the cost of all air circulation systems—cleanroom systems and others—is directly proportional to the air flow rate. Creative design engineering in cleanroom technology is thus consciously directed towards minimizing the air flow rate, by:

- reducing the room areas where unidirectional air flow is maintained to the minimum required by the process technology;
- reducing the flow velocity in the unidirectional flow areas to the minimum compatible with (a) thermal up-currents and/or flow instabilities caused by process-related heat sources and (b) turbulence caused by movements;
- confining the areas protected with unidirectional flow with suitable plastic curtains or dividing walls, thus separating them physically from the external areas with reduced protection requirements;
- employing alternative design concepts which effectively separate the process from contamination and people (e.g. minienvironments such as SMIF boxes for microelectronics and isolator or barrier systems for pharmaceutical applications—see Chapters 3 and 4).

CLEANROOM SYSTEMS: CUSTOM-MADE PROTECTION SCHEMES

The design basis of any contamination control protection scheme is the process to be protected, i.e. the sequence of operations, the layout of the machinery and equipment, and the specific cleanliness requirements for processes, material handling and circulation areas. The more exactly they are known, the more specifically can the cleanroom concept be adapted to the situation. Open-minded collaboration between the designer and supplier of the cleanroom system with the end user is absolutely essential if an optimal compromise between quality and flexibility on the one hand, and investment and running cost on the other hand, is to be achieved.

If only very small areas have to be protected with clean air, e.g. in a research laboratory, clean benches can provide a simple, flexible and at the same time cost-effective solution. Such clean benches are plug-in units operating with air aspirated from the room and are thus independent of its air conditioning system.

If the protection needs become more extensive, it is advantageous to integrate the provisioning of clean air into a central air handling system. Not only can flexibility be designed into the systems solution, but very energy efficient concepts can be created. In this case, different alternatives for air distribution present themselves, ranging from (a) spot protection of stand-alone workstations, (b) clean tunnels and (c) complete unidirectional flow rooms of vast dimensions: the so-called ballrooms.

For optimum flexibility of adaptation to the layout of a given production situation, it is frequently useful to separate the two functions:

- air filtration;
- establishment of the unidirectional flow field.

The first step is filtration of the supply air to the required cleanliness level, by means of HEPA or ULPA filters. Unidirectional flow is then established independently, downstream of the filters for example, by means of special air distribution elements capable of establishing low-turbulence displacement flow. One of the different options available for this is the air distributing element according to Grunder and Setz (Grunder, 1980; Bruderer, 1983) consisting of a double screen of woven polyester fabric. The shape of the distributors can be exactly adapted to the geometry of the machinery to be protected. If a conventional filter ceiling were used, adaptation would never be so perfect, due to the restrictions imposed by the standardized dimensions of HEPA or ULPA filters, and higher air flow rates would result.

Let us illustrate the three basic concepts:

- spot protection
- linear protection
- area protection

with a number of examples from industrial and hospital practice (Schicht, 1988).

Spot Protection

As an example of this protection principle, Figure 7.1 shows a sterile filling operation in the pharmaceutical industry. The supply air comes from a central air handling system and passes through HEPA filters. It is then distributed above the critical filling area with the aid of a clean air distributor, as described before. This produces unidirectional flow in a vertical downwards direction and ensures the requisite air cleanliness in accordance with Class 100 (ISO Class 5) in the core zone where the product and the primary containers into which it is being filled are openly exposed to the atmosphere. Plastic curtains effectively shield the aseptic core from its surroundings. These surroundings are maintained at the required air cleanliness level by simple spill over of air from the aseptic core, i.e. with no additional costs.

Figure 7.2 shows a typical operating room for surgical interventions requiring highly aseptic conditions, i.e. an environment of the wound and instrument table that is practically devoid of microorganisms. The HEPA filtered clean air is distributed directly above the operating area, so that displacement flow in a vertical downward direction, characterized by a very uniform field of velocity, is established. It has been found that a flow velocity of 25–30 cm/s is sufficient in this case. Therefore, both reducing the area where unidirectional air flow is maintained, and reducing the flow velocity will minimize the air flow rate.

Depending on the geometry of the areas to be protected and the type of manipulations being performed, it is necessary to select between vertical or horizontal, or in exceptional cases even diagonal, air flow. An example of spot protection with horizontal air flow is given in Figure 7.3, which shows a workstation in a micro-electronics research laboratory: by opting for horizontal air flow the required ambient conditions in respect of cleanliness, temperature and humidity in the area surrounding the complexly shaped electron-beam unit can easily be achieved—in spite of the large amount of heat emitted by the control and computing units visible on the left.

FIGURE 7.1. Spot protection of sterile filling operation in the pharmaceutical industry by vertical unidirectional flow produced by a clean air distributing element.

Linear Protection

A particularly widespread application of the principle of linear protection is to be found in the *clean tunnel* (Figure 7.4), which is quite commonly encountered in the micro-electronics industry. In this case the process equipment and apparatus is normally arranged in two rows along the side walls, behind which service and maintenance areas are provided. The establishment of unidirectional flow is brought as close as possible to the process equipment to be protected. The HEPA filters are provided with supply air from a central air handling system and are installed upstream

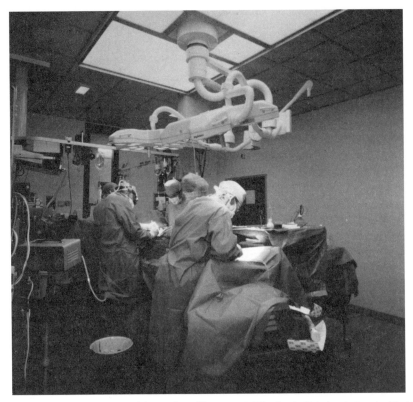

FIGURE 7.2. Operating theatre with an air ceiling panel. The basic lighting is located behind the air distributor made of a translucent synthetic fabric.

from the clean air distributing elements at ceiling level. This permits the arrangement of the fittings for lighting and any air ionization equipment upstream of the distributor, so that they do not interfere in any way with the flow field. The return air is normally extracted through openings in the side walls of each tunnel module into the maintenance area behind, from where it recirculates back into the central air handling system.

The main advantages of these clean tunnels with vertical flow are the following:

- the air cleanliness requirements can be differentiated (high standards for process zones, slightly reduced requirements for the circulation areas and corridors, considerably reduced requirements for the service and maintenance areas);
- minimum extension of the zones with unidirectional air flow (due to the interspersing of work and maintenance areas with each other);
- extraction of the return air laterally into the maintenance areas (this means that complicated floors and extensive underfloor cavities for conveying return air are avoided);
- low operating costs as a result of minimized air flow rates;
- high operational safety and simple subdivision into fire sections.

FIGURE 7.3. Spot protection by means of horizontal air flow in the area surrounding an electron-beam apparatus.

As drawbacks, the following must be noted:

- very careful and detailed planning is necessary to define the layout of the facility, space requirements for equipment and material flow requirements;
- reduced flexibility to accommodate changing operational requirements, equipment substitution and substantial layout alterations.

The direction of air flow also has to be selected according to the circumstances of each individual case when linear object protection is involved. For example, when assembly stations are to be protected—Figure 7.5 shows the assembly and quality testing of disc drives in a factory in the Far East—it is frequently preferable to opt for horizontal unidirectional flow directed towards the operators, because this can most effectively neutralize the particle emission from the personnel and eventual conveyor lines.

Area Protection in Extensive Cleanrooms

It is not always possible to fix the sequence of work steps in such a way that the principles of spot and linear protection, which are particularly attractive from the point of view of purchasing and operating costs, can be applied. There is then no alternative but to have recourse to the large-scale, whole-room solution.

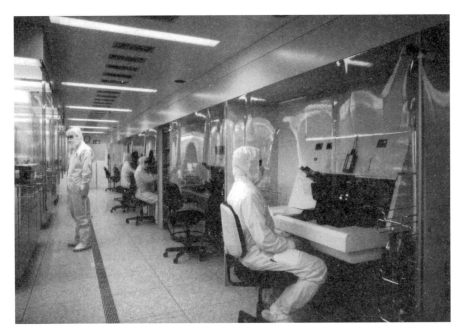

FIGURE 7.4. Production area designed in accordance with the tunnel principle.

A situation which requires the use of large cleanroom concepts is, for example, the assembly of space technology equipment, in which extremely high reliability and functional certainty are naturally prime considerations. The principle of horizontal air flow was selected for the room shown in Figure 7.6. This means that the critical work processes have to be concentrated immediately downstream of the air distribution wall, since work-related secondary contamination must be expected farther back in the room.

Particularly high air cleanliness requirements are to be met in the semiconductor industry. Although sometimes substituted by highly structured layouts based on the minienvironment principle (see Chapter 3), extensive ballroom-type cleanrooms continue to be designed and built in considerable numbers for Very large Scale Integration (VLSI) semiconductor manufacture. These rooms often have a floor area of several thousand square metres. When rooms of such dimensions and the correspondingly stringent requirements with regard to air cleanliness are involved, the utilization of the principle of vertical unidirectional flow is the only practical approach: supply air is introduced through large-area ULPA filter ceilings and the return air is removed through floor openings (Figure 7.7).

The most notable advantages of these large-area cleanroom zones are:

- maximum flexibility in locating the equipment and the partition walls;
- easy adaptation of the production room to changing production requirements or alterations in layout;
- simple conceptual design when standardized filter ceiling systems are used, since these permit very neat connection with wall systems of the same modular dimensions;

- simple air handling systems with constant supply air temperature; the subdivision of the room into zones for control purposes is often not necessary.

However, other, less attractive features of these large cleanroom systems should not be overlooked:

- the high air flow rates and the resultant high energy costs for air circulation as a consequence of the vast extension of these rooms, which are operated throughout on the principle of low-turbulence displacement flow;
- safety risks as a result of the limited possibility of subdivision into fire sections;
- the air cleanliness requirements can only be differentiated to a limited extent;
- extensive (and expensive) building volume above, adjacent to, and below the useful cleanroom area has to be provided purely for air circulation purposes.

Evaluation

Depending on the field of application, one or other of the concepts described here is preferable, with spot protection being predominant in research buildings, hospitals, the pharmaceutical industry as well as in small-scale production, while linear or ballroom-type area protection systems are more appropriate for extensive manufacturing and assembly plants. The choice between the different alternatives has to be made individually on a case-by-case basis and take into account a full appreciation of the specific

FIGURE 7.5. Linear protection with horizontal air flow for an assembly line for *Winchester* disc drives.

characteristics of each object. However, the concept adopted may not always have to be sharply defined: the solution may be to select a large-area clean zone for one part of a building and to use the tunnel principle or spot protection for another part.

OPTIMIZATION OF ENERGY CONSUMPTION IN CLEANROOM SYSTEMS

The different measures described so far aim above all at restricting the volumetric air flow rate to a minimum, and are effective in reducing both the investment and the operating cost. However, numerous other steps can be taken to reduce the energy consumption of a cleanroom system. Some of them may require additional investment, but as the total energy consumption during the system's useful life cycle may well sum to two or three times the investment cost—possibly even more—this may prove money well spent. Particularly good results are achieved by using a combination of measures which integrate a number of the individual steps listed below to form a harmonized whole (Schicht, 1982; Patel et al., 1991):

- variable outside air/recirculated air proportion in order to reduce the need for mechanical chilling by utilizing the principle of free cooling;
- use of waste heat for heating purposes;
- recovery of heat from the exhaust air with the aid of recuperation and reverse cycle heat pumps;

FIGURE 7.6. Assembly and testing of space vehicle components in a cleanroom with horizontal displacement flow, established by a clean air distributing element which fills the entire wall area in the background of the picture.

- minimal air humidification;
- appropriate system and zone subdivision, in line with the required room air conditions;
- application of the variable air volume principle whenever suitable;
- reduction of the air flow rate outside production hours;
- selection of system components combining high efficiency with low pressure drop;
- minimization of flow resistance in the air circulation system through low velocities, smooth transitions, no abrupt velocity changes, static regain where feasible, and other measures.

In addition to these possibilities, a whole series of further savings can be achieved, when planning air conditioning and cleanroom systems, by exhaustive use of the energy management potential of modern automatic control systems, especially when, by making use of micro-processor and personal computer technology, they are upgraded into a complete building management system.

EXAMPLE OF A FULLY INTEGRATED ENERGY CONCEPT

In the example given below an illustration will be provided of the way in which the energy-saving measures listed above can be integrated to form an overall solution. The premises in question are those of a research institute specializing in the development of application-specific semiconductors.

A central plant with constant air flow rate was chosen for the air conditioning of the

FIGURE 7.7. Cleanroom with vertical unidirectional flow, with a large-area filter ceiling and floor openings for the return air.

multi-storey building. At the entrance of each storey, the HEPA filters are grouped centrally in easily accessible compact filter boxes according to zones, so that monitoring and maintenance work can be carried out, without need to enter the research rooms. An air distribution system, which is designed so that later modifications can be made without difficulty, conveys the supply air to custom-designed clean air distribution elements. The size and arrangement of the distribution elements, the direction of air flow and the air flow velocity are exactly tailored to the individual requirements of each workstation. These are kept under clean air conditions of Class 100 (ISO Class 5). In some cases the apparatus is protected by the use of horizontal unidirectional flow, in others by vertical unidirectional flow.

The workstations are thus isolated from the surroundings by the use of the principle of spot protection. The remaining room areas of the laboratory are air conditioned merely by spill-over flow from the clean zones, and additional supply air devices have not been necessary. This allows both the desired room air conditions to e maintained and an air cleanliness corresponding to cleanliness Class 10 000 (ISO Class 7) to be ensured—at no additional cost as far as air engineering is concerned, as a first measure towards minimizing the air flow rate.

As a second measure with the same objective, the velocity of the air emerging from the clean air distributing elements was set individually within the range 0.25–0.4 m/s and is subsequently kept constant by means of automatic air volume control devices.

Minimization both of the spatial extent of the area protected by unidirectional airflow and its velocity as well as the use of air spill-over for the areas not protected with unidirectional flow are therefore used to keep down the air flow rate to the very minimum possible.

The air handling equipment is operated at only 70% of its rating. The reduction in the pressure loss and the higher degree of efficiency of the heat exchangers as a result of the lower air velocity more than compensate for the somewhat higher purchase costs of the units. Outside working hours the air flow rate is reduced to 50% of the nominal value, with the result that the energy consumption of the system for conveying the air decreases to less than 25%.

Air circulation is by means of high-performance fans with an efficiency at the point of operation of over 80%. This lowers the energy costs in comparison with a solution based on clean work benches even further, since that type of concept would require numerous small, decentralized fans with a relatively low efficiency.

Heat is recovered in three ways: the heat in the exhaust air is transmitted to the incoming outside air by means of finned heat exchangers; the cooling water, removing the waste heat from the equipment in the laboratory rooms, is used to further warm the entering outside air; and the waste heat of the water chillers is used to produce hot water for the sanitary installations and for the low-temperature heating system of the offices, and to supply the necessary heat for the multi-zone reheaters of the air conditioning system.

SOME SEMI-QUANTITATIVE COST INDICATIONS

After discussing some of the design options available for optimizing the investment and operational costs of cleanroom systems, some comments on cost levels may be indicated.

Although some data material concerning absolute cost levels can be found in the literature (e.g. Barnett et al., 1995; Lynn et al, 1993; Rakoczy, 1987), it seems wiser to refrain from such indications here. After all, absolute cost data are quickly outdated due to inflation. Furthermore, they vary considerably from country to country due to fluctuating exchange rates, different taxation practices, levels of salary and salary-related costs, interest rates, custom duties, insurance costs (especially where protection against product liability risks has to be provided) and many other factors. In addition, each cleanroom installation is an individual job, integrated into all sorts of different buildings, with considerable layout variation, different component mixes and quality standards. Some concepts are rather repetitive, others providing individual protection schemes for each workstation and therefore a higher engineering content. Permissible oscillations of control parameters, such as temperature and humidity, in respect of the mean values may be rather generous or very restricted, with important repercussions on the sophistication of the control system. Then there are the effects of scale and the influences of the mode of operation: running all the time at design capacity or with reduced air flow rates outside working hours, with 100% outside air or with maximum air recirculation, to name just a few of the relevant parameters influencing the investment and operational cost level.

Therefore, all the indications that follow can only serve as a very rough guideline. They can never replace individual job costing in all its detail.

Only in the field of micro-electronics are cleanroom concepts sufficiently uniform that a meaningful cost comparison between objects of different air cleanliness classes becomes feasible. Data developed from such a well-harmonized point of departure have been reported (Lynn et al., 1993). That study presents the cost data of a complete facility. The cleanrooms therefore are only one of many cost elements, as the study considers the general building costs, structural costs, the cleanrooms with all their elements and features, the mechanical systems comprising air circulation, heating and cooling, fire protection and other subsystems, and finally the electrical installations— only process equipment is excluded. Judged under these circumstances, and taking a Class 100 000 facility as point of reference:

- a class 1000 (ISO Class 6) clean facility would cost 1.45 times as much;
- a class 100 (ISO Class 5) facility 1.6 times as much;
- a class 1–10 (ISO Class 3 or 4) facility 1.75 times as much.

Unfortunately, all such studies differ widely in their data base, and any comparison between them is, as a rule, not very meaningful. To illustrate this effect, a second set of cost data is presented below (Rakoczy, 1987) which only takes the air circulation system of the clean facility into account. This set of cost data comprises the outside and recirculated air handling units and the fans, heat exchangers, humidifiers, sound attenuators, dampers and thermal insulation incorporated into them, all supply and return air ductwork, preliminary and final filtration, the complete chilled water circuit with refrigeration machines, cooling towers, pumps and all piping, the complete automatic control system and all necessary electrical wiring and switching material with the corresponding protection fuses. They do not include, however, partitions, floors and ceilings (other than filter ceilings), lighting, fire protection and other safety equipment, distribution networks for process fluids, automatic handling installations and they also exclude, of course, all process equipment.

In order to give some indications of installation costs, a simple, straightforward Class 10 000 (ISO Class 7) cleanroom system is taken as the point of reference. Comparing the cost per square metre

- a sophisticated Class 10 000 (ISO Class 7) cleanroom may cost five times as much;
- a unidirectional air flow system for Class 100 (ISO Class 5) twenty to thirty times as much;
- and a unidirectional air flow system for Class 1 (ISO Class 3) thirty to fifty times as much.

Looking at energy costs, a Class 10 000 (ISO Class 7) cleanroom system will cost five to ten times as much to run as a conventional air conditioning system for an office block, and a Class 100 (ISO Class 5) system fifty times as much,

How do the individual cost factors sum up into total cost? The cost split does not differ very much between turbulent mixing flow and unidirectional air flow systems: if we compare Class 10 000 (ISO Class 7) and Class 100 (ISO Class 5) cleanroom systems, then (Rakoczy, 1987)

- energy costs amount to 65%–75% of the total annual cost;
- capital costs, i.e. interest plus depreciation, to 15%–25%;
- and maintenance costs to 10% in both cases.

In order to reduce costs, therefore, energy costs merit particular attention. This is a point to be stressed particularly as it is disregarded in many investment decisions! The most pronounced potential contribution in order to reduce energy costs is air recirculation, with its fringe benefit of parallel reductions of investment cost (due to reduced chiller capacity requirements) and of maintenance cost (due to increased life expectancy of the HEPA or ULPA filters). In comparison with an installation running on 100% outside air, total cost may be reduced, due to air recirculation, by as much (Rakoczy, 1987) as

- a factor of two in case of a Class 10 000 (ISO Class 7) system;'
- and by a factor of three in case of Class 100 (ISO Class 5) system.

The total cost economy is less pronounced if regenerative or recuperative heat recovery systems have to be employed instead of simple air recirculation.

The Cost Impact of Minienvironments and Isolators

Although cleanroom concepts making use of minienvironments and isolators for separating the process from contamination and people are inherently more complex, their use may lead to very attractive cost savings. These result, on one hand, from the small space volumina occupied by the minienvironments themselves, and on the other hand from the fact that they are frequently able to be placed in an environment with quite a moderate air cleanliness classification: in the electronics industry, for example, Class 1 (ISO Class 3) minienvironments can be integrated into a Class 100 000 (ISO Class 7) cleanroom. This, in turn, leads to simplified garment requirements for the operators.

A 25%–40% reduction in cleanroom construction cost is reported (e.g. Barnett et al., 1995) when comparing a class 1 (ISO Class 3) ballroom-type facility for semi-conductor manufacturing with another one utilizing a minienvironment approach. This is, however, largely offset by cost increases for the process tools, so that the investment cost saving for the entire, fully equipped facility will be reduced to a mere 2%–4%. Most significant, however, is the reduction in running costs: the total air flow rate amounts to only one-third and the cooling load to only two-thirds of that required for a ballroom-type concept. For a typical facility of 3700 m^2 floor area, total annual savings of US$800 000 in the energy bill are reported, plus an additional saving of US$1.7 million a year as a consequence of the less rigorous garmenting concepts.

These economies, however, appear almost insignificant in comparison with other financial benefits claimed for minienvironments as applied in the microelectronics industry, and which are said to result from an accelerated ramp-up, i.e. the time required for the facility to achieve full production status, and from improved process yields (O'Reilly and Rhine, 1995; Livingston et al., 1995).

Similar economies may be obtained when employing isolators for aseptic processing operations in the pharmaceutical industry (Ruffieux, 1995). He reports a recent case where the use of isolators permitted a 10%–15% reduction in investment cost in com-parison with a conventional cleanroom concept. In addition, he expects a 40% reduc-tion in operating costs—due mainly to reduced air circulation as well as to less sophisticated garmenting schemes.

CONCLUDING REMARKS

The wide variation in the cost level of cleanroom systems opens up a considerable space for technical creativity. In defining optimal problem solutions, detailed knowledge of the available design options and the relevant selection criteria is required from the design engineer. Consistent systematic thinking and strict consideration of the specific circumstances of the intended utilization of the building, thorough familiari-zation with each problem posed, and therefore with the mentality and the concerns of cleanroom users embracing a wide variety of professional disciplines, are other impor-tant and necessary qualifications. In the end, only the design achieving, in a strongly competitive environment, the best compromise between effectiveness and functional safety on the one hand, and investment and operating cost on the other hand—or, in other words, the design with the best value/cost relationship—will win the customer's preference.

ACKNOWLEDGEMENT

Figures 7.1–7.7 are reproduced by permission of Zellweger Luwa Ltd, Uster, Switzerland.

REFERENCES

Barnett, W. J., Castrucci, P., Schneider, R. K. and Williams, M. M. (1995). 'Cost reduction of cleanrooms with 100% minienvironments', *Proceedings 41st IES Annual Technical Meeting*, Anaheim, 30 April–5 May. Institute of Environmental Sciences, Mt. Prospect, Illinois, USA, pp. 134–141.

Bruderer, J. (1983). 'An economic and efficient air distribution method for the establishment of a clean working environment', *Swiss Pharma*, **5**, 11a, 17–21.

Grunder, H. (1980). 'Ueber ein neues System der gezielten Anwendung von Laminar-flow (On a new concept of localised application of laminar-flow)', *Reinraumtechnik*, **IV**, 7–10, Swiss Society for Contamination Control, Zurich, Switzerland.

Livingston, J., Art, D., Martin, R. and Traylor, F. M. (1995). 'Are minienvironments more economical for new construction?', *Microcontamination*, **March**, 33–39.

Lynn, Ch., Sears, M., Stanley, E. M. and Startt, S. (1993). 'Cleanroom facilities—Cost and budget development', *CleanRooms '93 Conference*, Boston, 22–25 March.

O'Reilly, H. and Rhine, B. (1995). 'The role of minienvironments in redefining microelectronic manufacturing economics', *Proceedings 41st IES Annual Technical Meeting*, Anaheim, 30 April–5 May. Institute of Environmental Sciences, Mt. Prospect, Illinois, pp. 130–133.

Patel, B., Greiner, J. and Huffman, T. R. (1991). 'Constructing a high-performance, energy-efficient cleanroom', *Microcontamination*, **February**, 29–32, 70–71.

Rakoczy, T. (1987). 'Die Kosten von Reinraumanlagen und ihre wirtschaftliche Optimierung (The cost of clean room systems and their economic optimization)', *Reinraumtechnik*, **VIII**, 39–43, Swiss Society for Contamination Control, Küsnacht, Switzerland.

Ruffieux, P. (1995). 'Application examples of isolator technology II—Preparation of solutions and filling in sterile drug manufacturing', *Pharma Technologie Journal, Isolator Technology in Aseptic Manufacturing*, **16**, 1, 43–54.

Schicht, H. H. (1982). 'Application of energy-saving clean room systems in research and manufacture of micro-electronic components', *Proceedings of the 6th International Symposium on Contamination Control*, Japanese Air Cleaning Association, Tokyo, pp. 87–90.

Schicht, H. H. (1988). 'Engineering of cleanroom systems—General design principles', *Swiss Contamination Control*, **1**, 6, 15–20; **2**, 1, 27–39

8 High Efficiency Air Filtration

S. D. KLOCKE and W. WHYTE

INTRODUCTION

The supply of air to a cleanroom must be filtered to ensure the removal of particles that would contaminate the process being carried out in that room. Until the early 1980s, that air was filtered with High Efficiency Particulate Air (HEPA) filters, which were the most efficient air filters available. Today, HEPA filters are still used in many types of cleanrooms but one cleanroom application, the production of integrated circuits, has evolved to a level where more efficient filters are required. These are known as Ultra Low Penetration Air (ULPA) filters.

In cleanrooms, high efficiency filters are used for the dual purpose of removing small particles and, in unidirectional flow cleanrooms, straightening the airflow. The arrangement and spacing of high efficiency filters, as well as the velocity of air, affect both the concentration of airborne particles and the formation of turbulent zones and pathways in which particles can accumulate and migrate throughout the cleanroom. The combination of a high efficiency filter and a fan only initiates the unidirectional flow process. A balance of the entire airflow path is required to ensure good unidirectional flow.

Although various cleanroom standards define the level of airborne particles in various classes of cleanrooms, e.g. 'ISO Class 6 or Class 1000' different arrangements and types of HEPA and ULPA filters are required to provide cleanrooms for different industries. The two largest industrial cleanroom applications are in the semiconductor and pharmaceutical industries. However, cleanrooms are now used in biotechnology, medical devices, food processing, disk drive and other industries. The semiconductor industry uses ULPA filters in fabrication areas for the production of submicron geometries, whereas HEPA filters are sufficient to remove bacterial and inert particles that can cause problems in pharmaceutical and other cleanrooms.

It is generally accepted that for cleanrooms of ISO 6 (Class 1000) and higher, HEPA filters are sufficient to meet the room classification, and traditional ventilation techniques, such as the use of terminal filter units or filters installed in the air supply ducting, are adequate. For ISO 5 (Class 100), HEPA filters should completely cover the ceiling, supplying unidirectional flow down through the cleanroom. For ISO 4 (Class 10) or lower, ULPA filters should be used in a unidirectional flow cleanroom.

Cleanroom Design. Edited by W. Whyte © 1999 John Wiley & Sons Ltd

THE CONSTRUCTION OF HIGH EFFICIENCY FILTERS

High efficiency filters are usually constructed in two ways, i.e. deep pleated or mini-pleated. Both methods are used to ensure that a large surface area of filter paper is compactly and safely assembled into a frame so that there is no leakage of unfiltered air through it.

In a deep-pleated filter, rolls of filter paper are folded back and forward, side by side, either in 6 in. (15 cm) or 12 in. (30 cm) lengths. To allow the air to pass through the paper and give the filter strength, a crinkled sheet of aluminum foil is often used as a separator. This pack of filter media and separators is then assembled and glued into a frame, which could be of a plastic, wood or metal construction. A cross-section of such a traditional construction is shown in Figure 8.1. Deep-pleated construction is most often used for filters in conventional ventilation systems where the application velocity is usually 250–500 ft/min (1.25–2.5 m/s).

An alternative method of deep pleating is to corrugate the filter medium during the media production process and fold the media directly into a pleated pack so that the corrugations support the pleats as well as space them out. Such a construction is illustrated in Figure 8.2.

More recently, high efficiency filters have also become available in a mini-pleated form. Aluminum separators are not used in this method of construction, but the paper medium is folded over ribbons, glued strings, or raised dimples in the media and

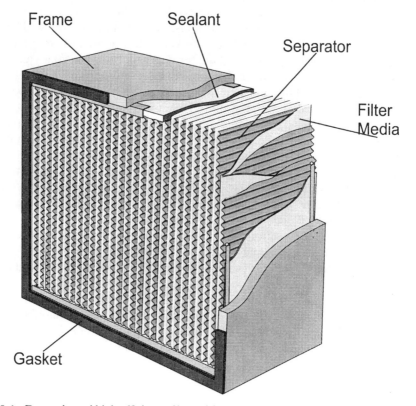

FIGURE 8.1. Deep-pleated high efficiency filter with separators.

assembled into a frame. This method of assembly allows six to eight pleats per inch (2.5 cm) compared with approximately two to three pleats per inch (2.5 cm) found in the deep-pleated filters. The mini-pleated filter contains much more medium for the same surface area so that these filters can be made more compact. Mini-pleated construction is the most widely used method of construction for a unidirectional flow cleanroom because the larger media area yields a lower pressure drop than a deep-pleated construction for common unidirectional flow velocities of 70–100 ft/min (0.35–0.5 m/s). Such methods of construction are shown in Figure 8.3 and 8.4.

HEPA FILTERS

A HEPA filter, is defined by its particle removal efficiency and its pressure drop at a rated airflow. A HEPA filter is defined as having a minimum efficiency in removing small particles (approximately equal to 0.3 μm) from air of 99.97% (i.e. only three out of 10 000 particles, 0.3 μm in size, can penetrate through the filter).

The traditional size of a deep-pleated type of HEPA filter is 2 ft × 2 ft × 12 in. (0.6 m × 0.6 m × 0.3 m), which has a rated flow of 1000 ft^3/min (0.47 m^3/s), at a maximum pressure of 1 inch water gauge (250 Pa) will have between 170 ft^2 (15.9 m^2) and 275 ft^2 (25.5 m^2) of filtering media. Dividing the airflow by the area of the media gives an air velocity of between 3.6 ft/min (1.8 cm/s) and 5.9 ft/min (3.0 cm/s) at the rated flow. This

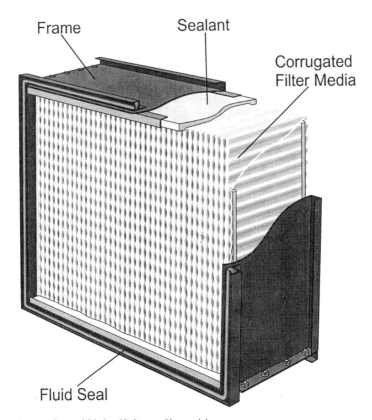

FIGURE 8.2. Deep-pleated high efficiency filter without separators.

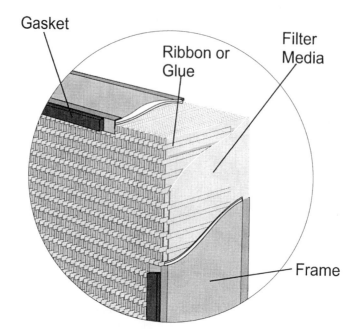

FIGURE 8.3. Section of a mini-pleated filter with separators.

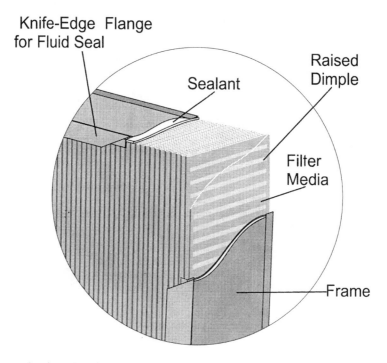

FIGURE 8.4. Section of a mini-pleated filter without separators.

velocity if very important, because it determines the removal efficiency of the filter medium and if the air velocity is increased or decreased the efficiency will change. It is possible, by increasing the amount of filtering medium in a filter, not only to decrease the pressure drop across it but also to increase its efficiency.

ULPA FILTERS

The category of ULPA filter was created to define filters that have efficiencies higher than those of standard HEPA filters. An ULPA filter will have an efficiency greater than 99.999% against 0.1–0.2 μm particles. These filters are constructed, and will function, in the same way as a HEPA filter. They differ in that the filter medium used has a higher proportion of smaller fibres and the pressure drop is slightly higher. For a filter with the same amount of medium, an ULPA filter will have a higher resistance than a HEPA filter. Because of the higher efficiency of removal of smaller particles, the methods used for testing HEPA filters are not appropriate and other methods using laser optical particle counters or condensation nuclei counters are employed.

PARTICLE REMOVAL MECHANISMS

A high efficiency filter is designed to remove particles of about 2 μm and smaller. Much less expensive pre-filters can be used to remove larger particles but these will not be discussed in this chapter. A high efficiency filter medium is made of glass fibres ranging in diameter from as little as 0.1 μm up to 10 μm, with spaces between fibres quite often very much larger than the particles captured. These fibres criss-cross randomly throughout the depth of the filter medium and do not give a controlled pore size. A photomicrograph of a sample of high efficiency filter medium is shown in Figure 8.5. As an aerosol moves through the filter paper, the entrained particles bump into the fibres, or onto other particles that are already stuck to the fibres. When a particle bumps into either a fibre or a particle, a variety of strong forces, such as Van der Waal's, are established between the captured particle and the fibre or particle that has captured it; these retain the particle.

The three most important mechanisms in the removal of small particles in a filter medium are impaction, diffusion and interception; of lesser importance is sieving or straining. These are shown diagrammatically in Figure 8.6. It may be assumed that electrostatic effects are unimportant in high efficiency filters.

In the process of capture by *impaction*, particles large enough (i.e. with enough mass) to have sufficient momentum will leave the gas stream and strike a fibre as the gas turns around a fibre. In the process of capture by *diffusion*, small particles (i.e. those without sufficient mass to leave the gas stream on their own) move about randomly because they are constantly bombarded by other small particles and by the molecules of the gas in which they are suspended. This random motion causes the smaller particles to touch the fibres or previously captured particles. If a particle strikes a fibre as it passes it, i.e. tangentially, it will be captured or retained, this mechanism being *interception*. The final mechanism of filtration, which is known as *sieving* or straining, occurs when the spaces between the fibres are smaller than the particles that are being captured.

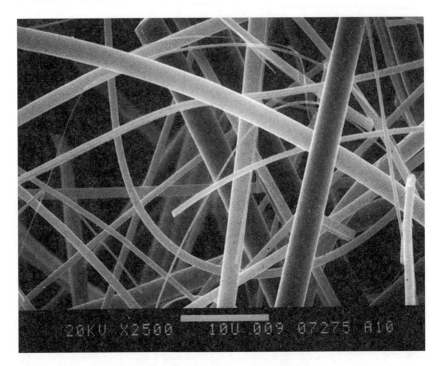

FIGURE 8.5. Photomicrograph of high efficiency filter medium.

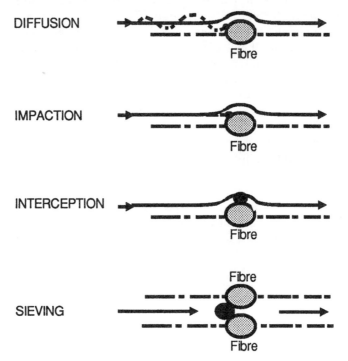

FIGURE 8.6. Particle removal mechanisms.

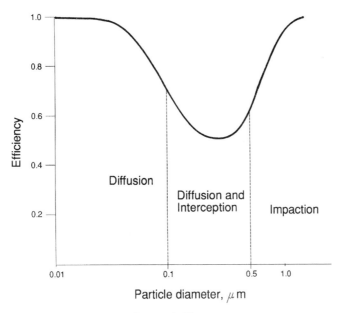

FIGURE 8.7. Classical efficiency curve for an air filter.

A high efficiency filter is dependent on three of the mechanisms described above, the larger particles being removed by inertial impaction, the medium size by direct interception and the smallest by diffusion. This concept is alluded to in Figure 8.7, which gives the classical efficiency curve for a HEPA-type filter showing that the minimum efficiency of such a filter is for a particle of about 0.3 μm and particles smaller or larger will be removed more efficiently.

This is a rather simplistic approach, because it is known that the minimum efficiency (or the maximum penetrating particle size) is dependent on a number of variables. These are:

1. Particle density
2. Velocity and mean free path of the particle
3. Thickness of the filtering medium
4. Velocity, pressure and temperature of the gas in which the particle is aerosolized
5. Sizing and distribution of fibres within the medium

Because penetration is a function of so many variables, it may be best not to think in terms of a single most penetrating particle size. Most aerosols are not homogeneous. Aerosols in a cleanroom may contain great varieties of particles, from skin flakes to silicon. The velocity through the filter may vary as can the filter fibre diameter and packing density of the fibres. Drs Lui and Lee, and their colleagues at the University of Minnesota, have investigated such variables and on the basis of their work, some of which is shown in Figures 8.8 and 8.9, it may be best to think in terms of a most penetrating particle size range from 0.1 μm to less than 1.0 μm, rather than a single most penetrating size.

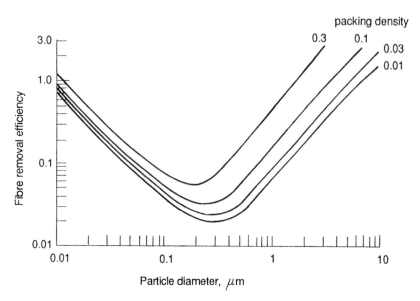

FIGURE 8.8. Fibre removal efficiency of particles in relation to packing density and particle diameter. Fibre diameter 3 μm and air velocity 10 cm/s.

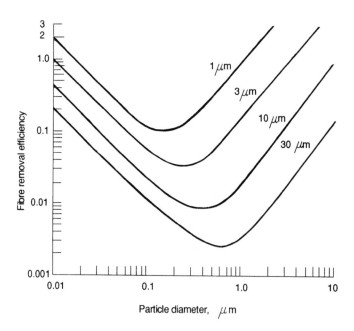

FIGURE 8.9. Fibre removal efficiency of particles in relation to diameter of fibres in filter. Air velocity 10 cm/s and fibre packing density 0.1.

THE HIGH EFFICIENCY FILTER AS STRAIGHTENER

Most high efficiency filters are constructed so that the filter paper is arranged in a large number of parallel pleats. The pleats are very narrow and deep. A typical 2 ft × 4 ft × 6 in (0.6 m × 1.2 m × 0.15 m) high efficiency filter contains 140–190 such pleats. These pleats straighten the direction of the air as it flows through the filter. In addition, the resistance of the high efficiency filter paper to airflow is reasonably uniform, which means that approximately equal volumes of air move simultaneously through each pleat of the filter. Those two factors, the uniform resistance of the filter paper and the large number of parallel pleats, cause the air to flow uniformly downstream of the high efficiency filter face. It is this uniformity of flow that initiates the phenomenon of unidirectional flow, causing small particles to move in the direction of the path of the flow.

The use of a high efficiency filter does not guarantee unidirectional flow. Other features must be controlled to maintain unidirectional flow through and out of a cleanroom. Moreover, some high efficiency filters are constructed with V-banks of pleated media and filters of this type are not intended for unidirectional flow applications.

THE TESTING OF HIGH EFFICIENCY FILTERS

High efficiency filters are tested to measure their efficiency against test particles. There are a number of tests used for such purposes. They are outlined in the following subsections.

United States Military-Standard 282 (Mil-Std 282)

This test method is used to measure the performance of HEPA filters and is described in the US Army specification Mil-Std 282, 'DOP Smoke Penetration and Air Resistance of Filters'. This specification describes the operation of a special instrument known as a Q-107 penetrometer which is shown in Figure 8.10. The penetrometer measures the aerosol penetration through a filter and its initial resistance to airflow at a specified flow rate.

The test begins with the manufacture of particles that are nearly homogeneous in size (with a count median diameter of about 0.2 μm). To test a typical filter at 1000 ft³/min (0.47 m³/s) on the Q-107 penetrometer, outside air is drawn into a duct at 1200 ft³/min (0.57 m³/s) and then divided into three parallel ducts at 85 ft³/min (0.04 m³/s), 265 ft³/min (0.125 m³/s), and 850 ft³/min (0.4 m³/s). As shown in Figure 8.10, the top duct contains banks of heaters and a reservoir to contain the liquid material to be aerosolized. In the original development of this test method, the aerosol material of choice was dioctylphthalate (DOP). However, DOP has been listed as a suspected carcinogen and, more recently, other materials such as polyalphaolefin (PAO) and dioctylsebacate (DOS) have been substituted. These materials have no known adverse health effects and give essentially identical test results when the photometer in the instrument is calibrated with the same material. The centre duct contains a cooling coil and a bank of heating elements. The air passing through the top duct is heated to approximately 365°F (185°C), and is then impinged through an orifice onto the liquid aerosol material in the reservoir. This causes the liquid to evaporate and it is carried forward to the confluence of the top and centre ducts where it is quenched by the cooler air from the centre duct. The nominal 0.3 μm particle size is controlled here by

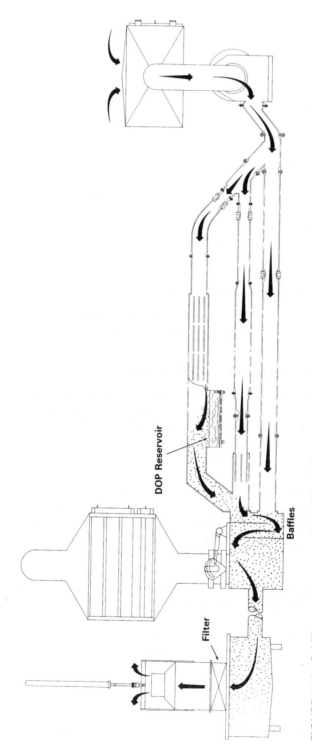

FIGURE 8.10. Q-107 penetrometer used in Mil-Std 282.

maintaining a temperature of around 72°F (22°C) and by increasing or decreasing the temperature it is possible to increase or decrease the particle size.

The combined airflow from the upper two ducts is then mixed with the remaining 850 ft^3/min (0.40 m^3/s) from the bottom duct. A series of baffles mixes the aerosol (smoke) thoroughly into the air stream so that the aerosol is uniform, prior to challenging the filter. A similar set of baffles is located on the exhaust side of the filter being tested so as to thoroughly mix the effluent. An upstream sample is taken and, when the aerosol concentration reading is between 80 mg/m^3 and 100 mg/m^3, that value is accepted as a 100% challenge. A reading (percentage concentration) is now taken downstream of the filter (downstream of the baffles so that any leakage is thoroughly mixed into the effluent) and is compared with the upstream value. This is read as a percentage penetration and when subtracted from 100%, the filter's efficiency is obtained.

Sodium Flame Test (Eurovent 4/4 and British Standard 3928)

This European test method uses an aerosol of particles of sodium chloride mainly within the size range of 0.02 μm–2 μm and having a mass median size of 0.6 μm. This test aerosol is used to determine the filter's efficiency.

The test apparatus is shown in Figure 8.11. The test aerosol of sodium chloride is generated by atomizing a sodium chloride solution and evaporating the water. Most of the larger droplets are removed by baffles placed around the atomizers and the remaining cloud is carried down the ducting by the main airflow. The large droplets escaping the baffles are deposited at the entrance of the ducting and are drained away. Thorough mixing of the sodium chloride aerosol is ensured by a circular baffle situated centrally in the ducting downstream of the spray box. The relative humidity of the air and the dimensions of the ducting are such that the aerosol is substantially dry before it reaches the mixing baffle and all water is evaporated by the time it reaches the filter.

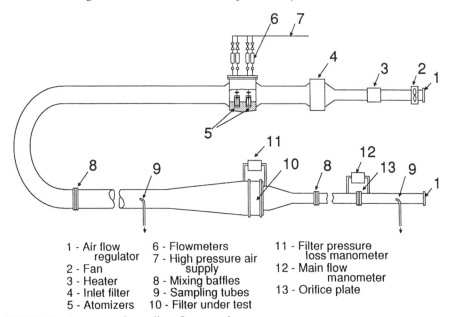

1 - Air flow regulator	6 - Flowmeters	11 - Filter pressure loss manometer
2 - Fan	7 - High pressure air supply	12 - Main flow manometer
3 - Heater	8 - Mixing baffles	13 - Orifice plate
4 - Inlet filter	9 - Sampling tubes	
5 - Atomizers	10 - Filter under test	

FIGURE 8.11. Apparatus for sodium flame testing.

After passing through the filter under test, the air flows through the downstream ducting, and any entrained sodium chloride is uniformly dispersed over the cross-section of the duct by a further baffle and an orifice plate. The orifice plate is provided with a manometer for the measurement of the airflow. A continuous sample of the air is drawn off downstream of the orifice plate through a sampling tube situated coaxially in the duct and facing upstream. Metered flows of sample air and the methane, necessary for combustion, are mixed before passing into the photometer burner. In the methane gas flame, the sodium chloride particles present in the sample air will emit yellow light, characteristic of sodium. This light, after passing through a suitable optical system, falls upon the sensitive face of a photoconductive cadmium sulphide cell. The cell forms part of a zero reading on a microamperemeter when no sodium chloride is present, i.e. when the flame is 'neutral'. The sodium chloride passing through a filter under test produces an out-of-balance reading on the microamperemeter, which can be converted to a penetration value by prior calibration of the instrument with sodium chloride of a known concentration. This penetration value is expressed as an efficiency, that value being used to calculate the filter's efficiency.

The size distribution of the sodium chloride test aerosol is similar to the Mil-Std 282 test and these two tests give similar results; however, these two methods have not been proven to be equivalent. Shown below in Table 8.1 is a comparison of the two methods employed on the same filter.

TABLE 8.1. Comparison of filter efficiencies;
Mil-Std 282 and sodium flame tests.

Mil-Std 282 (%)	Sodium flame (%)
0.003	0.001
0.01	0.05
5.0	4.2

Institute of Environmental Sciences and Technology (IEST) Recommended Practice 'Testing ULPA Filters'

The IEST has developed a Recommended Practice for testing ULPA filters (IES-RP-CC-007; 'Testing ULPA Filters'). An optical particle counter with a laser light source can measure particles smaller than 0.3 μm with convenience, accuracy and reproducibility. Condensation nuclei counters can extend the lower range. A system using these particle counters can be used to test ULPA filters for efficiency.

The system consists of inlet filters, a blower, an aerosol generator, a duct and plenum for mounting the filter to be tested, a sampling system for both the upstream and downstream aerosol concentrations, a flow measurement and pressure measurement device, and the optical particle counters. Air is drawn into the system by the blower and filtered with HEPA or ULPA filters. The method requires that the concentration of the background aerosol be lowered to less than 5% of the total test aerosol. The test aerosol is injected into the test airflow by any suitable means of generation. The generator could be a compressed air atomizer type such as the Laskin Nozzle for liquid aerosol materials. More recently, ultrasonic humidifying equipment has been used to aerosolize polystyrene latex sphere solutions in water. The choice of the aerosol material is open to the user, but the material must meet certain optical properties that are listed in the method. The aerosol is typically polydispersed because the optical particle counters can count and size the aerosol as opposed to the monodispersed aerosol of the Q-107 penetrometer. The airflow is measured with either a calibrated orifice or nozzle or other flow-sensing device. The method allows the use of either a single particle counter to sample sequentially the aerosol from upstream and downstream, or two particle counters to sample simultaneously both concentrations. For either system, the particle counter draws in a known flow rate of air from both the upstream and downstream sides of the filter. Both samples are uniformly mixed by agitation or baffles in the duct. A dilution device permits sampling of the very high concentration of aerosol in the upstream challenge. The measured downstream concentration is divided by the upstream concentration and the dilution ratio, which yields the penetration for each size range of the particle counter. The penetration is subtracted from unity and multiplied by 100% to yield the efficiency. This test system provides the operator with a particle size efficiency in size ranges from around 0.07 μm to 3.0 μm. A computer can present the efficiency data in both tabulated and graphical form. This system far exceeds the sensitivity of the Q-107 penetrometer (used to test HEPA filters) and can therefore be used to verify the performance of ULPA filters. It may also be used to measure the efficiency of HEPA filters, but this particle count method should not be considered equivalent to the efficiency as measured by the Q-107 penetrometer.

European Standard (EN 1822)

At the time of writing, this standard is in its final phase of approval. It will be used for classifying both HEPA and ULPA filters and is based on the German DIN Standard 24 183. It gives a method for both testing the particle removal efficiency of filters and classifying them according to their efficiency.

The EN 1822 is available in five sections. These are as follows:

- Part 1: Classification, performance testing, marking
- Part 2: Aerosol production, measuring equipment, particle counting statistics
- Part 3: Testing flat sheet filter media
- Part 4: Testing filter elements for leaks (scan method)
- Part 5: Testing the efficiency of the filter element

An important departure of this test method from previous methods is the determination of the Most Penetrating Particle Size (MPPS) for the filter medium being tested and the determination of the removal efficiency of the filter at that particle size. As has been discussed in the 'particle removal mechanisms' section of this chapter, each filter has a particular particle size that will pass through the filter most easily, that size being determined by variables, such as the fibre diameter of the filter medium, air velocity and its packing density. It is logical, therefore, to test the filter at that most penetrating particle size. The MPPS is normally between 0.1 μm and 0.3 μm.

The first stage of this test method is to determine the MPPS of the flat sheet filter medium used in the filter. This is carried out at the face velocity that will correspond with that produced by the filter when working at its given volumetric flow rate. The efficiency of the complete filter element is then determined.

The complete filter element is placed in a test rig where the complete filter face is scanned to ascertain the penetration of particles of a size determined as the MPPS. The scanning is carried out to ascertain the local values of penetration caused by leakage through pinholes, etc. From the local values, the overall penetration and efficiency can also be calculated. As an alternative, the overall penetration of the most penetrating particles through the complete filter element can be carried out. This will be done using a fixed downstream probe.

The filter is then classified by its overall and local efficiency against its most penetrating particle. Shown in Table 8.2 is the proposed classification for HEPA and ULPA filters.

TABLE 8.2. Classification of filters according to the proposed EN 1822 scheme.

Filter class	Overall value efficiency (%)	Overall value penetration (%)	Local value efficiency (%)	Local value penetration (%)
H 10	85	15	–	–
H 11	95	5	–	–
H 12	99.5	0.5	–	–
H 13	99.95	0.05	99.75	0.25
H 14	99.995	0.005	99.975	0.025
U 15	99.999 5	0.0005	99.997 5	0.002 5
U 16	99.999 95	0.000 05	99.999 75	0.000 25
U 17	99.999 995	0.000 005	99.999 9	0.000 1

Probe (Scan) Testing of High Efficiency Filters

Air is supplied in a conventionally ventilated cleanroom through diffusers or terminal filters covering only a small portion of the room ceiling. The required overall efficiency of the filtration system is defined and some pinhole leaks in the filters can be tolerated, as long as they are not great enough to reduce significantly the overall efficiency of the filtration system and affect the required air cleanliness. This tolerance is possible because a small number of particles passing through leaks will be well mixed with the turbulent clean air,

This is not the case in unidirectional flow systems, as found in clean work stations and cleanrooms, where the high efficiency filters are located as a filter bank either in a ceiling or a wall located in close proximity to the process or product to be protected. To dilute a pinhole leak with the rest of the clean air passing through the filter, either a considerable distance or some method of agitating, such as a baffle, would be required. A baffle would be pointless in a unidirectional flow cleanroom. It is therefore possible that the product or process requiring particle-free air could be located directly downstream of a pinhole leak and hence be contaminated by particles.

Early researchers into cleanroom techniques realized the problem of pinhole leaks and developed a procedure to scan or probe the downstream face of a bank of filters in a unidirectional flow room, not only to locate pinhole leaks in the filter element, but to determine whether the filters were sealed to their mounting frames. This type of test method is discussed in the 'in-service tests' section at the end of this chapter. High efficiency filter manufacturers, confronted with the prospect of failing a field test that could locate defects that could escape detection in the overall efficiency test with the Q-107 penetrometer, began to manufacture probe filters destined for cleanrooms. In time, this additional test requirement became an industry standard and is described in IEST-RP-CC-001 'HEPA and ULPA Filters'. Although the IEST document recommends a specific procedure for each level of overall efficiency, the customer often specifies the procedure to be used in the factory so that it is the same or slightly less stringent than the procedure for the field test. Both photometer and particle counter procedures are defined in that document. These procedures are described in the next two sections.

Photometer Method   As shown in Figure 8.12, there are three major components used to perform the probe or scan test in the filter manufacturer's factory, i.e. the aerosol

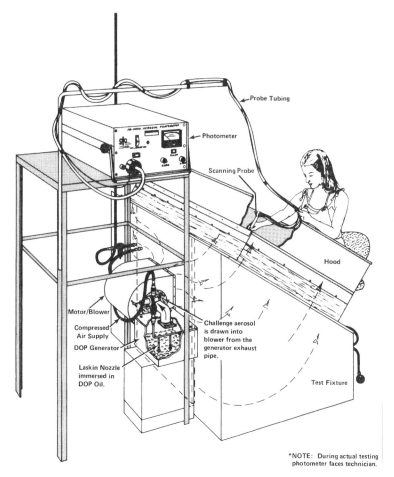

FIGURE 8.12. Probe-testing for pinhole leaks.

generator, the test box plenum with motor/blower and the photometer. The aerosol is most often generated using a Laskin nozzle generator in which compressed air at 20–25 psig (138–172 kPa) is forced into a liquid reservoir. The flow of compressed air causes the liquid passing through the Laskin nozzle to be fragmented into an aerosol.

Unlike the homogeneous, monodispersed aerosol produced by the Q-107 penetrometer, this aerosol is polydispersed. The aerosol size distribution varies slightly with the choice of liquid material, but when DOP liquid is used and compressed air pressures are as stated above, the distribution, by weight, is approximately:

- 99% less than 3 μm
- 95% less than 1.5 μm
- 92% less than 1 μm
- 50% less than 0.72 μm
- 25% less than 0.45 μm
- 10% less than 0.35 μm

Although test plenums vary somewhat in size and design, the arrangement shown in Figure 8.12 is typical. The essential purpose of the plenum is to mix the air with the oil upstream of the filter to provide a uniform challenge to the filter. An important feature of the test equipment is the hood or baffle that is located on the air-leaving side of the filter. This device prevents the entrainment of particles from the room air into the downstream face of the filter and is essential to obtain valid results. During the test, the filter is clamped in place between the hood and the test plenum. For some applications, the filter is mounted in the plenum, so that it is encapsulated by the plenum. This pressurizes and challenges the filter frame joints in addition to the media and media-to-frame bond. In older photometers, the operator set the needle of the meter to the zero point while holding the probe at the filter face and sampling the effluent from the filter. Current photometers contain their own filter for setting the zero reading. Next, an upstream reading is taken through an orifice in the plenum upstream of the filter. If the challenge is insufficient, an adjustment is made by increasing the air pressure into the generator and checking the upstream concentration reading until the correct limit is attained. The filter is now scanned for leaks at the perimeter and, using slightly overlapping strokes, over the face of the filter; the photometer will indicate leaks that are greater than 0.01% of the concentration before the filter.

Probe-tested filters can be erroneously described as 'zero probe' or '99.99' filters with the inference that they have a higher efficiency rating than the minimum overall efficiency of 99.97%. As described above, the procedure for probe-testing includes setting the meter at zero while sampling the effluent from the filter being tested. This procedure would therefore give the same results if a low efficiency filter were tested and is unrelated to overall efficiency. Some photometers are now equipped with HEPA filters that are used as reference filters, but there is no industry-wide specification requiring their use. The probe test is therefore a supplement to, not a substitute for, the overall efficiency test.

Optical Particle Counter Method   This method is essentially the same as the photometer method, except that a discrete optical particle counter is used as the sensing instrument. It is most often applied to ULPA filters but works equally well with HEPA filters. The aerosol is generated and introduced to the upstream side of the filter and the face of the filter is scanned, as before, with the probe using overlapping strokes. The upstream particle count concentration must be measured. This presents some difficulty as the high concentration on the upstream side is often beyond the capability of the particle counters used to scan downstream. Three techniques have been developed to solve this problem. The first is to construct and use an aerosol diluter when sampling the upstream concentration. The dilution ratio and stability must be determined prior to using it. IEST-RP-CC-007 describes the calibration of a diluter. Another is to use a different particle counter of a small enough sample flow rate so that it can sample directly from the upstream concentration. Lastly, a correlation between a photometer response and the particle count concentration can be developed for a specific set of test conditions.

The particle counter method is very sensitive and it is best to perform it inside a clean space in the factory environment, because any stray particles from outside the downstream side of the filter may be perceived as a leak and slow down the test by rechecking false leaks. The choice of the value of the particle count to indicate a leak is important. Using a count of one gives a high probability of false leaks and missing true leaks. It is therefore better to choose a count of 10 or more and increase the upstream concentration to keep the percentage leak the same.

The choice of aerosol for use in the particle counter method is mostly made according to customer requirements. In semiconductor cleanrooms, for example, outgassing from the filters is a critical concern and so oily liquid aerosols are not allowed. A common material is polystyrene latex, manufactured as spheres that can be purchased dispersed in water. The aerosol is usually generated using an ultrasonic humidifier. This creates a fine mist of a dilute solution of the spheres and water. The water evaporates leaving the spheres as the test aerosol. The aerosol size distribution is determined by the size of the spheres, which are available from submicron to several microns in diameter.

The semiconductor industry also requires tighter specifications for the size of acceptable pinhole leaks in ULPA filters and an acceptable leak rate much closer to the overall penetration of the ULPA filters. Current technology allows a leak rate of 0.001% to be tested economically although the proposed European standard is 0.0001% for the highest efficiency filters. This must be performed using automatic positioning (scanning) equipment to control scan speed rigidly, laser particle counters interfaced to a computer to monitor very low particle count rates and a cleanroom environment in which to test. These improvements ensure the accuracy of the test, provide a hard copy output from the test and, when run in a very clean environment, can be used to indicate leaks as small as 0.001% (compared with the 0.01% sensitivity of the photometer method). However, only ULPA filters with overall efficiencies of 99.9999%, or greater, are probe-tested to this level.

A HEPA filter that has a performance rating of, for example, 99.99% with the probe test, is one that has no pinhole, crack or imperfections showing an indicated leak greater than 0.01% of the upstream concentration at a specific location. This test should not be compared with the overall efficiency rating obtained with the Mil-Std 282 test, because there are so many differences, including the particle size distribution and concentration of the challenge.

Institute of Environmental Sciences and Technology (IEST) Recommended Practice 'HEPA and ULPA Filters'

The IES has published a Recommended Practice (IEST-RP-CC-001) on 'HEPA and ULPA Filters', which describes both suitable construction materials and performance levels for common applications of filters. The document also groups filters into six grades according to their construction. These are as follows:

- *Grade 1.* Requires filters to meet all of the material and design qualification tests listed in the US Army specification Mil-F-51068. Filters meeting these requirements are used in nuclear safety systems. The US Army has now deleted this specification and replaced it with a new consensus standard ASME AG-1 'Code on Nuclear Air and Gas Treatment' and may be used by specifiers in its place. Presumably, future revisions of the IEST document will correct this reference.

- *Grade 2.* Requires filters to meet the Underwriters' Laboratories Test (UL-586), necessitating fire-retarding construction to ensure the filter's ability to maintain most of its efficiency in the event of fire or when heated air approaches the filter in service. The UL-586 is performed as a design qualification test. The UL-586 test is required by Mil-F-51068 and ASME-AG-1 but there are many other requirements in those documents. The IES document classifies this grade as a filter meeting this one requirement. It is an appropriate filter for containment safety systems.

- *Grade 3.* Requires filters to meet the UL-900-Class 1 standard, and therefore requires the filter to be made of fire-retarding materials and emit a minimal amount of smoke or flame when exposed to flame or heated air in service. This ensures a certain level of safety to personnel occupying the space served by the filters. This is commonly specified for cleanroom HEPA and ULPA filters.

- *Grade 4.* Requires the filter to meet UL-900-Class 2 standard and the filter to be, therefore, of moderate fire-retarding construction. This allows some smoke to be emitted from the filter when exposed to flame and heated air. This is the most common construction level for general heating, ventilation and air conditioning (HVAC) systems.

- *Grade 5.* Requires the filter to be Factory Mutual Listed and be of fire-retarding construction, but there is no requirement regarding smoke emitted. The filter is tested installed in its supporting framework of ceiling grid, the ceiling grid and filter being qualified as a system. This is often specified for unidirectional flow cleanroom filters.

The performance of HEPA and ULPA filters is divided into six groups in this Recommended Practice. The integral efficiency is measured by either a Q-107 penetrometer or a particle counter system. It also lists recommended performance levels and test methods for a leak test. The grades are shown in Table 8.3.

Two-Flow Efficiency Testing and Encapsulation

Pinhole leaks in filter media will give greater penetration at lower velocities because the constriction of airflow through a pinhole is a function of the square of the air velocity, whereas the resistance to airflow through the filter media is close to a linear

TABLE 8.3. Grades of high efficiency filters.

	Penetration test		Scan test		Minimum
Type	Method	Aerosol	Method	Aerosol	Efficiency (%)
A	Mil-Std 282	DOP	None	None	99.97 @ 0.3 μm
B	Mil-Std 282	DOP 2-flow test	None	None	99.97 @ 0.3 μm
C	Mil-Std 282	DOP	Photometer	Polydisperse	99.99 @ 0.3 μm
D	Mil-Std 282	DOP	Photometer	Polydisperse	99.999 @ 0.3 μm
E	Mil-Std 282	DOP 2-flow test	Photometer	Polydisperse	99.97 @ 0.3 μm
F	IES-RP-CC-007	open	Particle counter	Polydisperse	99.999 @ 0.1–0.2 μm

Note: The aerosol listed for the penetration measurement is listed as DOP for the MIL-Std 282 test, but other materials are known to be satisfactory.

function. Therefore, at 20% of full test flow, a pinhole leak shows up as approximately 25 times greater, in proportion to total flow, than it does at full flow.

The development of acceptance criteria for cleanroom components resulted in a great awareness of the existence of pinhole leaks in high efficiency filters. Not wishing to commit additional time to prove-test high efficiency filters or to raise the purchase price for additional factory testing, a two-flow test was adopted where the HEPA filters were tested at the flow rate specified in ASME-AG-1 and again at 20% of that flow rate. The 20% flow test will detect any gross pinhole leakage escaping notice in the full flow test. The test is not effective in detecting all pinhole leaks nor does it enable the operator to locate them, but it has been found to be an effective device to improve overall filter performance.

At the time that the 20% flow test was added to the procedure, a second modification was also made. This was the addition of the encapsulation hood shown in Figure 8.13.

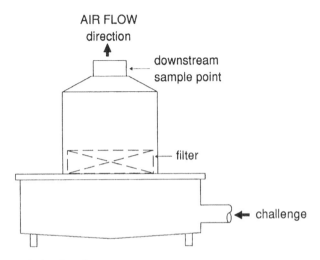

FIGURE 8.13. Encapsulation hood.

Previously, filters tested for efficiency had only the element and the frame tested. Experience has shown that HEPA filters can have frame leakage, caused primarily by improper sealing methods during manufacture, racking of the frame or leakage through the frame material. By enshrouding the entire filter, any leakage through the frame, joints or corners is included in the measurement of overall efficiency of the filter.

Currently, the nuclear industry in the United States specifies the two-flow leak test method. This is defined as a Type B filter in IEST-RP-CC-001. Some biohazard applications also specify the two-flow leak test and the probe test to be performed as an extra precaution. This is defined as a Type E filter in IEST-RP-CC-001.

FILTER HOUSINGS FOR HIGH EFFICIENCY FILTERS

When a high efficiency filter leaves the factory where it has been manufactured and tested, it will be fit for the purpose required. If it has been properly packed and transported and installed by personnel who are familiar with the delicate nature of the filter media, then the filter's integrity should be maintained. However, to ensure that

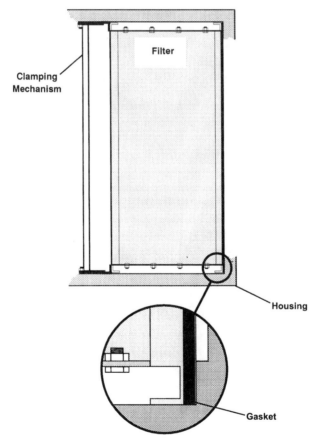

FIGURE 8.14. Traditional neoprene gasket sealing method shown in a duct-mounted filter housing.

there is no ingress of unfiltered air into the cleanroom, the filter must be fitted into a well-designed housing. The housing must be of sound construction and particular attention must be paid to the method of housing/filter sealing. Traditionally, neoprene rubber gaskets are fitted to the filter and these press down onto a flat face of the housing to prevent leakage (Figure 8.14). This method can be successful, but distortion of the filters or frames, as supplied or when tightening up, as well as poor gaskets, can cause leakage. Better designed housings overcome these problems.

Admissible Air Leakage Seal

In this method, the base of the filter housing is fitted with a special gasket, as shown in Figure 8.15. The filter is then fitted into the housing and a pressure test is carried out. The air channel in the gasket is pumped up to a given pressure (usually 2000 Pa) and the amount of air that flows into the gasket to constantly achieve this pressure is noted. A maximum amount of leakage is set and any problems with the housing/filter gasket can be found easily.

Fluid Seal

Figures 8.16, 8.17 and 8.18 show typical applications of this method. Figure 8.16 shows a filter fitted with a channel for a fluid seal, which mates into a housing used in a conventional ventilating system. This housing is typically located in the supply duct system. A knife edge is fitted to the filter housing, which mates into the channel of sealant. The fluid will flow around the knife edge to give a perfect seal, yet will not flow out of the channel. No leakage can occur through this filter-to-housing seal.

Figure 8.17 shows a housing that is mounted in the ceiling of a conventional ventilation system or a mixed flow cleanroom. A filter is installed from the room side and contains the fluid seal in a channel around the perimeter of the filter.

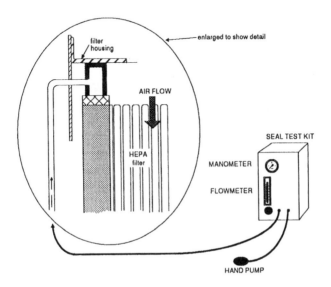

FIGURE 8.15. Air leakage test method.

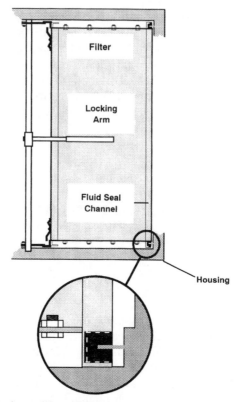

FIGURE 8.16. Filter with channel for a fluid seal. Duct-mounted filter housing used in conventionally ventilated cleanroom.

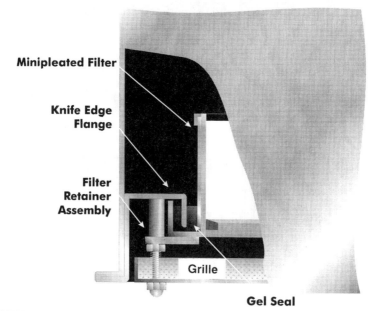

FIGURE 8.17. Filter with channel for a fluid seal. Filter used in conventionally ventilated cleanroom and installed from room.

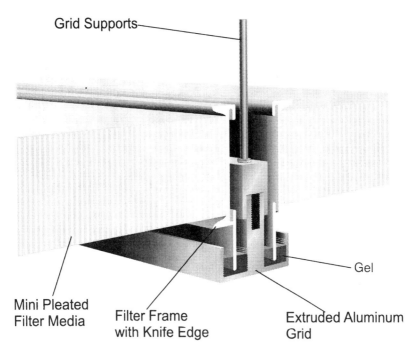

Grid Supports

Gel

Mini Pleated
Filter Media

Filter Frame
with Knife Edge

Extruded Aluminum
Grid

FIGURE 8.18. Ceiling grid with channel for a fluid seal. System used in a unidirectional flow cleanroom.

Figure 8.18 shows a ceiling grid that forms a channel for the fluid seal in which filters are placed. This system would be used in a unidirectional flow cleanroom.

The term 'fluid seal' is used almost universally. However, it usually refers to a material known as a gel that actually is a very soft solid. The material is either a silicone or urethane compound that is poured into the channel on the filter, or mounting frame or ceiling grid and that cures to its final properties in place. The gels have the advantage of being able to facilitate the removal and replacement of the filter from its housing with near 100% reliability and very low clamping force.

IN-SERVICE TESTS FOR HIGH EFFICIENCY FILTERS

Stringent factory tests for testing high efficiency filters have resulted from the requirements of both the nuclear and cleanroom industries. Experience has demonstrated that high efficiency filters that have passed these tests do not always arrive at their destination without mishap. Damage can occur during shipping and handling and, once installed, the filter-to-housing seal or leaks in the housing can contribute to a loss of efficiency. Consequently, cleanroom users may require validation of the in-service performance of both high efficiency filters and their supporting frameworks. The testing of high efficiency filters in cleanrooms is carried out by probe- or scan-testing procedures to ensure that the filters are correctly installed. Filters that have first been factory probe-tested should also be tested in-service in a manner similar to the factory test. The installed filter and its supporting framework are challenged by introducing a particle challenge aerosol upstream of the filter bank and by scanning the downstream face of the bank with either a photometer or a particle counter. The instrument choice is dependent on the application and the aerosol material. Pinhole leaks in the filter media and filter-to-frame leaks are then identified and repaired or the filter replaced.

Synthetic and natural oils in aerosolized form have been extremely useful for many years for probe-testing high efficiency filters in cleanrooms to locate defects in filters and their frames and housings. These materials are used because they are inexpensive and easy to aerosolize using a small container, a Laskin nozzle and compressed air; for greater quantities, machines are available that will generate thermally large amounts of the aerosol. The cold-generated oily liquid aerosols provide a difficult challenge for any high efficiency filter bank, and being oil droplets they have a minimal impact on filter loading. Dioctylphthalate (DOP) has been used for over 35 years but the more recent concerns about its health effects have resulted in declining use. However, tests carried out at the Research Triangle Institute, North Carolina, indicate that the amount of DOP that off-gasses from a fully tested HEPA filter, i.e. one that has been tested with DOP for both efficiency and defects, is indistinguishable from the amount of DOP in the earth's atmosphere. Other oils with no known health hazard are now used and their choice will vary and is dependent upon the country in which the testing is carried out. Dioctylsebacate (DOS), Polyalphaolefin (PAO), Diethylhexylsebacate (DEHS) and Shell Odina oil are popular choices.

All of the above oils can be aerosolized using a Laskin nozzle, which aspirates the liquid with compressed air, or they can be produced as a smoke from a thermal generator. The photometer should be calibrated with the same material as is used for testing. This is especially true when the photometer is used to measure an absolute mass concentration of the aerosol and not just the relative concentration downstream compared with the upstream.

An optical particle counter can be used in lieu of the photometer and provides increased sensitivity so that less aerosol material may be used. Solid or liquid aerosol materials have been successfully used with a particle counter. Oil concentrations lower than 0.008 mg/l of air can be used to test filters in-place with a particle counter but with a photometer a concentration of about 10 mg/l will be required. Where there has been a concern about the outgassing of materials from the tested filters, the oily liquid aerosol materials have been replaced with aerosolized polystyrene latex spheres, which are available in a wide range of submicron sizes. However, in most cases, the generators for this aerosol are not capable of a high enough output to use a photometer. Although the particle counter method is generally considered more sensitive, no definitive study has been published on this subject. The particle counter method also has the drawback of requiring the operator to discern a concentration or particle count rate from discrete counts measured in time by the counter as the filter is scanned.

Oily test aerosols are suspected of contributing to the contamination of the product and/or process in semiconductor and disk drive cleanrooms. In pharmaceutical or other cleanrooms, they are unlikely to harm the product. Off-gassing of the aerosol used to test the high efficiency filters has been blamed for defects in the silicon wafers. Today, the construction materials for ULPA filters used in these cleanrooms are scrutinized quite closely. In such a situation, testing of the filter in the factory and in-place can be carried out using particle counters with atmospheric dust that can be supplemented, if necessary, with an appropriate aerosol.

When only atmospheric dust is used to challenge the filter being probe-tested, the challenge usually varies over time and varies with the environmental conditions, so that the downstream criteria used with a particle counter must be altered accordingly. Therefore, the desire for a constant aerosol challenge using a material known to be safe for the semiconductor process has resulted in the use of polystyrene latex spheres (PSLs). These spheres, suspended in water, are available from several manufacturers. They can be aerosolized easily using ultrasonic humidification equipment. This equipment has been shown to be easily adjustable, repeatable and quite economical compared with oily liquid aerosol materials. The PSL spheres are available in a wide range of monodispersed solutions from around 0.1 μm to 1 μm, so the size distribution of the aerosol is tailored easily to the range of interest.

By not electing to test the filters in-service, the owner and/or designer generally issues a less stringent procurement specification, with the result that the housings could be of mediocre quality because they are not designed to pass an in-service test. If it should be decided at a later time to upgrade the filter installation and add in-service testing, modification or replacement of housings may be necessary. A major reason for the failure of these filter banks, assuming that undamaged filters have been installed, is the filter-to-housing seal. Bypass leakage has long been a principal cause of improperly installed filters. This can be compounded by housing leaks in welds, or caulking, or by poor quality workmanship by the installer of the housings. If HEPA filters are not probe-tested at the factory, then a certain percentage will fail the in-place probe test.

Standard cleanroom design criteria call for unidirectional flow rooms and clean air devices to have a face velocity of up to 90 ft/min ± 20 ft/min (0.46 ± 0.1 m/s). A considerable amount of air therefore passes through the filter. It is not difficult in smaller unidirectional flow systems to generate sufficient amounts of challenge aerosol. However, as the size of the system increases, it becomes increasingly more difficult to challenge the entire bank simultaneously. One alternative method of testing is to isolate adjacent sections and test the bank section by section. Where this is not practical, the filters can be tested at the job-site prior to installation, using a similar test rig to that used at the factory for probe-testing. This test establishes that there has been no damage to the filters incurred in transit, or during unpacking and handling. Following this test, the filters are installed immediately under close supervision. When all filters have been installed, the seals between the filter and the supporting framework itself and the perimeter of the filter bank are scanned for bypass leakage.

Considering the time that it takes to scan manually a large high efficiency filter wall or ceiling bank, there is a requirement for a less time-consuming test that allows a cleanroom ceiling to be more quickly monitored. Such a test is that using multiple particle counters. Multiple sampling probes can be arranged on top of a trolley and placed just under the filter so that as the trolley is moved across the cleanroom the probes will sample not one point of the filter, but its entire length. This method is quicker and will therefore allow lower amounts of particle challenge to be collected on the filters.

FILTER STANDARDS

BS 3928 (1969). *Method for Sodium Flame Test for Air Filters*. British Standards Institution, UK.
Mil-Std 282. *DOP-Smoke Penetration and Air Resistance of Filters*. US Government Printing Office, Washington, USA.

European Standard: EN1822. *High Efficiency Air Filters (HEPA and ULPA)*. European Committee for Standardization, B1050 Brussels.

Eurovent 4/4 (1976). *Sodium Chloride Aerosol Test for Filters using Flame Photometric Technique*. European Committee of the Constructors of Air Handling Equipment, Wein, Austria.

Recommended Practice IEST-RP-CC-001. *HEPA and ULPA Filters*. Institute of Environmental Sciences and Technology, USA.

DIN 1946-4 (1989). *Raumlufttechnik; Raumlufttechnische Anlagen in Krankenhäusern (Heating, Ventilation and Air Conditioning; HVAC Systems in Hospitals)*. Beuth Verlag GmbH, D-10772 Berlin, Germany.

Recommended Practice IEST-RP-CC-007. *Testing ULPA Filters*. Institute of Environmental Sciences and Technology, USA.

Recommended Practice IEST-RP-CC-006. *Testing Cleanrooms*. Institute of Environmental Sciences and Technology, USA.

DIN 24184 (1990). *Typprufung von Schwebstoffiltern; Prüfung mit Paraffinölnebel als Prüaeosol (Type Testing of High Efficiency Particulate Air Filters; Using Paraffin Oil Mist as Test Aerosol*. Beuth Verlag GmbH, D-10772 Berlin, Germany.

ACKNOWLEDGEMENTS

Figures 8.1–8.4, 8.10, 8.12–8.14 and 8.16–8.18 are reproduced by permission of Flanders Filters. Figures 8.8 and 8.9 are reproduced by permission of Dr B Lui. Figure 8.11 is reproduced by permission of the British Standards Institution. Figure 8.5 is reproduced by permission of Evanite Fiber Corporation.

9 Construction Materials and Surface Finishes for Cleanrooms

E. C. SIRCH

INTRODUCTION

The cleanliness of the surfaces of cleanrooms can have an essential influence on the cleanliness of products manufactured in the room. Materials used in the construction of a cleanroom should therefore be chosen to ensure that they do not release particles, or other contaminants, that will contaminate the product. To achieve this the materials should be (a) easy to clean and, where necessary, resistant to water, detergents and disinfectants; (b) durable, non-shedding and chemically inert; and (c) when required, antistatic. The construction of the cleanroom also requires that the materials used will ensure an airtight structure. These properties should be retained throughout the life of the cleanroom. To ensure that these requirements are achieved, the type of materials and features of construction described below should be considered.

In this chapter all inner surfaces of a cleanroom are considered, with the exception of the air supply unit and its final high efficiency filters. Also considered are the adjacent personnel changing areas and material airlocks, the doorways, and any decontamination equipment and conveyors connected to the cleanroom.

GENERAL CONSIDERATIONS

All components used in the construction of a cleanroom should comply with the relevant local regulations concerning fire protection, thermal insulation, noise control, site electrical transition resistance, dynamic forces connected with vibrations, and the static load. These building regulations are covered by national laws.

The relevant national guidelines and rules for conditions in the work-place must also be considered. These will vary from country to country but could contain requirements for:

- avoiding glaring surfaces;
- ensuring visibility from the work-place to the area outside the factory;
- providing escape routes in case of fire or emergency situations.

Cleanroom Design. Edited by W. Whyte © 1999 John Wiley & Sons Ltd

Some national and internal guidelines, especially in the healthcare industry, also influence the design criteria for cleanrooms. For example, in the European Union's Guide to Good Manufacturing Practice for pharmaceuticals (see Chapter 2 of this book) the requirements for the quality of surface cleanliness required are as follows:

- no horizontal surfaces above the operation level;
- surfaces to be smooth, impervious and without sharp edges, free of pores, abrasion resistant, unbroken, easy to clean, as well as resistant to cleaning agents and disinfectants;
- antistatic, if necessary;
- double glazed windows and doors, to be flat and smooth;
- proper wall or ceiling openings for light fixtures and services to the cleanroom;
- joints sealed with plastic or elastomers, e.g. silicon or silicon rubber.

The specified physical, chemical and constructional properties of cleanroom components must conform to four main requirements to ensure the correct performance of a cleanroom. These are as follows:

1. Functionality
2. Durability
3. Cleanability
4. Maintainability

The correct selection of construction materials and finishes depends on first establishing the four main performance criteria that are required for the cleanroom to be constructed and then selecting the proper finish and substrate materials which comply with the building regulations and relevant guidelines mentioned above. The construction material and finish requirements, as well as the detailed architectural design, are dependent on the cleanroom standard required, the higher the standard of the room the higher the quality of materials that should be specified. In selecting appropriate material finishes and methods of construction, the need to protect against contamination should take into consideration the degradation and wear of the material, or finish, used in the construction materials. High-quality finishes may provide the best technical specification and the best aesthetic conditions, but the higher investment cost may not be necessary to meet the required performance. Cost-effectiveness should be judged not just only on the initial cost, but upon a life cycle analysis which takes account of the running costs for repair, and the impact of renovation or alterations.

Figure 9.1 lays out the hierarchy of all the requirements which should be considered in the selection materials used in cleanrooms.

PERFORMANCE CRITERIA OF CONSTRUCTION MATERIALS AND SURFACES

As discussed above, the choice of construction materials for a cleanroom is dependent on their functionality, durability, cleanability, and maintainability.

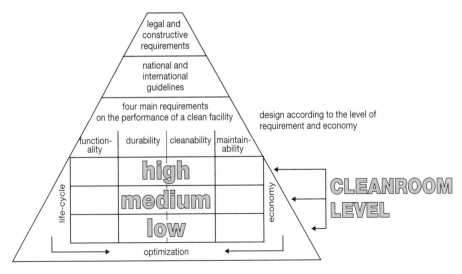

FIGURE 9.1. Hierarchy of requirements for construction materials used in cleanrooms.

Functionality

Specialized functions, independent or unrelated to the contamination risk of the product, are often required in a finish system. For example, the finish, substrate, and design detail are usually required to be sufficiently airtight to provide proper air pressurization and defined humidity conditions. They may also be required to maintain specific conductivity or anti-static requirements. In certain instances, anti-microbial aspects required in pharmaceutical or food production, or the outgassing properties of surfaces in the semiconductor industry, may be a concern in selecting the correct material and finish systems.

The accumulation of an electrostatic charge, and the subsequent electrostatic discharge, is a problem in nearly all cleanrooms. It may give an unacceptable risk of (a) an explosion (in the presence of powders or gases), (b) device damage (e.g. damage to electronic or optical components), or (c) contamination through the attraction of particles to charged surfaces. Where the above electrostatic risks cause concern, materials used in the construction of installations should neither generate nor hold a significant static charge. This anti-static requirement will be specific to each application, and should be clearly specified by the purchaser. Certain processes may require particular conditions, in terms of environmental humidity, to minimize the generation and retention of an electrostatic charge. To minimize electrostatic problems, the relative humidity should not be less than 50% RH for continuous operation, and should never drop below 30% RH for more than a few hours.

To protect components sensitive to electrostatic discharge the resistance to earth for the entire structure should be in the range of 10^4 to 10^7 Ω. Care should be taken to protect the personnel against the risk of electrocution. Earthing should be considered, with a site transition resistance of 5×10^4 Ω. The 'ideal' range of resistance is therefore between the site transition resistance of 5×10^4 Ω and the mass resistance of 10^7 Ω.

The required electrical characteristics for flooring are valid for the entire structure, or composition of materials used as a floor, and these should be measured from time to time to monitor potential loss of performance through ageing. The accumulated surface charge should not exceed 2 kV.

Durability

All finishes should be able to withstand the traffic they bear and the necessity for maintenance, repair or replacement should be within acceptable limits. Durability is most important in areas where friability or degradation of finishes can serve as a source of product contamination. Finished floors must therefore be capable of withstanding the rolling wheel load of fork trucks, pallet jacks, tanks, etc. where these are used.

Durability considerations should include chemical resistance to process ingredients and cleaning and disinfectant agents which come in contact with the surface. Other products, or materials used in production, may have particular agents which will stain surfaces, with certain finish systems being more susceptible to attack than others.

Cleanability

Walls, floors, and ceilings in cleanrooms should be constructed so that the primary surfaces are accessible for cleaning (primary surfaces are the first solid surfaces which would be contacted when moving away from an exposed product). In a cleanroom, primary surfaces generally include the walls, floors, ceilings and doors, the room side of air diffusers but not the duct side, floor drain, etc.

Because walls, floors, or ceilings require washing or wiping down on a frequent basis, consideration in the selection of materials should evaluate their properties to withstand this level of cleaning. This is particularly important where disinfectants are used, as ideally they should remain in contact with the surface for several minutes. If the cleanroom is washed down, the benefit of coved intersection details and integral floor bases should be considered. Horizontal surfaces are acceptable if they are within reach for cleaning.

Surfaces in cleanrooms should promote cleanliness by providing unbroken, non-porous, smooth and, if necessary, anti-static surfaces and details which minimize areas of dust collection. Coved wall-to-wall, wall-to-ceiling, or wall-to-floor details, along with integral floor bases, should be considered as details which enhance cleanability. Right-angled corners between walls and floors are required when cleaning by machines.

Maintainability

Finish materials must retain the performance consistent with the level of protection required. This may require regular maintenance procedures. Alterations or repairs should be able to restore the qualities of the materials to the original condition. In general, better finishes will cause less disruption and repair effort, and serve better the architectural and aesthetic requirements and all needs of the process.

CONSIDERATIONS FOR SPECIFIC COMPONENTS

There are many acceptable methods and materials for constructing cleanrooms ranging from *in situ* construction to fully prefabricated site-assembled systems. The basic options are summarized as follows:

- *In situ* construction: Wet or dry construction with applied surface finish.
- *In situ* assembly: Pre-finished engineered components, modular pre-finished composite panel system.

Combinations of these basic construction options can also be used. In what follows only *in situ* assembly systems for walls and ceilings and wet construction systems for floors are illustrated because they are the most common systems.

EXAMPLES OF MATERIALS AND FEATURES OF CONSTRUCTION

This section gives guidance on the selection of acceptable architectural details, materials, and finishes for use in cleanrooms. The components described are materials and components used commonly in cleanroom construction. In general, the use of higher quality components is an acceptable option. However, in all cases the appropriate selection should be based upon the level of protection required, and performance criteria as outlined in an earlier section in this chapter.

To ensure that dirt does not accumulate and that surfaces can be easily cleaned, the horizontal surfaces and ledges in a cleanroom must be minimized. The following should therefore be considered when constructing a cleanroom:

- flush glazed windows;
- flush mounted door frames;
- flush mounted light fittings and switches;
- enclosed piping chases.

Given below is a range of materials that should be considered as materials suitable for the construction of a cleanroom. No suggestions are made, at this stage, as to which of these materials give a higher performance.

A primary property of the materials used in cleanroom construction is their ability to shed few particles. Materials that are considered to be *non-shedding* and used in cleanroom construction are as follows:

- stainless steel;
- powder-coated sheet metal (or anodized in the case of aluminium);
- sealed concrete;
- plastic sheets, hot welded;
- non-shrinking coatings of plastic materials;
- ceramics;
- glass.

Materials considered to be *resistant to abrasion and against cleaning agents* are finishes that are unbroken and free of pores and are as follows:

- stainless steel surfaces;
- baked finishes;
- epoxy and resin paints and finishes;
- ceramic surfaces;
- powder-coated finishes;
- sealed concrete.

Surfaces that are considered to be *cleanable* are as follows:

- smooth finished stainless steel;
- coated metal (or anodized in the case of aluminium);
- plastic polymers;
- glass;
- smooth finished glazed tiles.

To ensure that airborne contamination does not penetrate the cleanroom from adjacent areas of higher airborne contamination and to ensure that the cleanroom structure is airtight, it is necessary to use the correct type of openings where the structure is penetrated. Examples of where *suitable openings* are required are as follows:

- sealed light fixtures;
- sealed piping and wiring penetrations (especially sprinkler heads);
- sealed materials of construction joints in walls/ceilings/floors;
- sealed filter frames;
- door and entryway seals.

CLEANROOM COMPONENTS FOR GOOD SURFACE CLEANLINESS AND LOW DETERIORATION

Floor Systems

Effectiveness against deterioration caused by wear and ageing is the main criteria for a good floor system used in a cleanroom. To obtain effective floor systems they should be built following the construction principles shown in Figure 9.2.

The general requirement for the durability of a good floor cover are:

- a well crafted screed with the best quality materials, good workmanship and with a sufficient ageing period;
- consideration of the slots required in the construction i.e. expansion joints, building joints;
- a construction that will stand up to static and dynamic stress;
- and, if necessary, a shunt for static discharge.

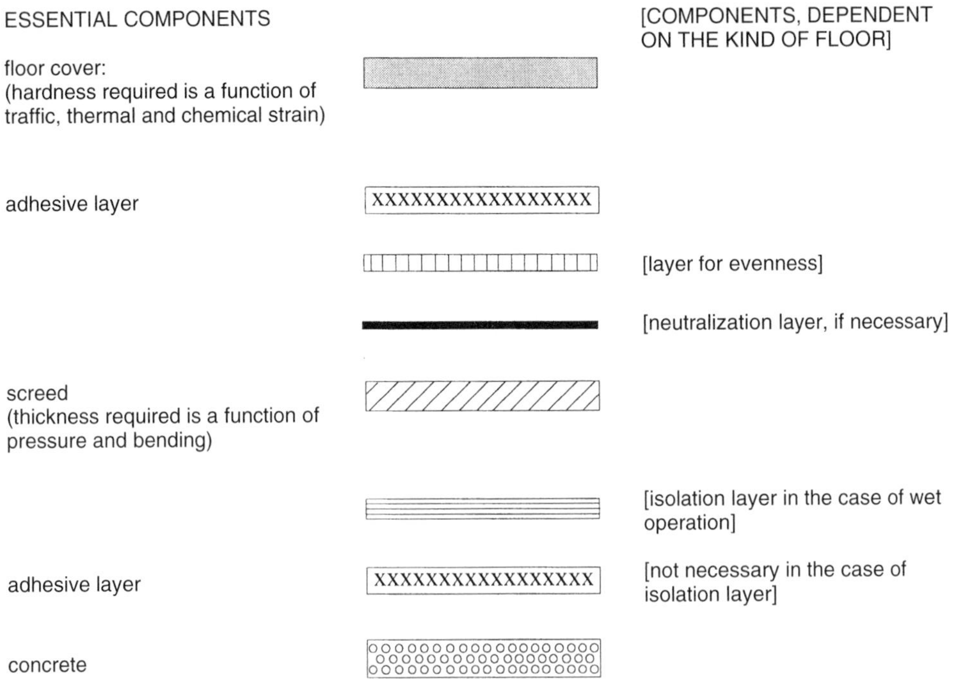

ESSENTIAL COMPONENTS

[COMPONENTS, DEPENDENT
ON THE KIND OF FLOOR]

floor cover:
(hardness required is a function of
traffic, thermal and chemical strain)

adhesive layer

XXXXXXXXXXXXXXXX

[layer for evenness]

[neutralization layer, if necessary]

screed
(thickness required is a function of
pressure and bending)

[isolation layer in the case of wet
operation]

adhesive layer

XXXXXXXXXXXXXXXX

[not necessary in the case of
isolation layer]

concrete

FIGURE 9.2. Transverse section through a suitable floor system. [ ] = shows optional components.

Figures 9.3–9.8, which show cross-sections through suitable cleanroom floor systems, are examples of methods that can be used for different cleanliness classes and different applications.

Wall Systems

Cleanroom walls are constructed either by an *in situ* construction method (similar to that used in ordinary building methods), or by *in situ* assembly where pre-built components are joined together to form a cleanroom.

In situ construction   In this method of construction, walls made of blocks or metal stud partitions are used as the basic framework of the rooms. Because utility services pass through the walls, metal stud partitions are often more adaptable. If block walls are used, then to ensure a tight, hard and smooth finish they should be skimmed with a layer of plaster and epoxy painted, or economically finished by lining with 12 mm gypsum plaster board, which is taped and epoxy painted. Both these methods are more suitable for lower standard cleanrooms. A glass fibre reinforced epoxy finish is more suitable for a higher standard of cleanrooms.

In situ assembly   In this method the walls are prefabricated and assembled on site. The walls are normally free standing and hence have to be of sufficient strength and

HIGH CLASS FINISH FLOOR SYSTEM

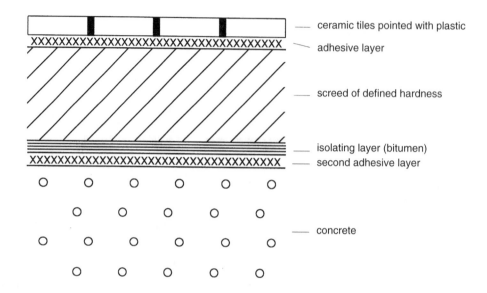

FIGURE 9.3. Tiled floor for wet operations.

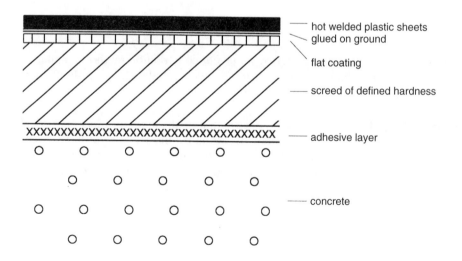

FIGURE 9.4. Hot welded plastic sheets on screed.

HIGH CLASS FINISH FLOOR SYSTEMS SUITABLE FOR MINIMIZING ELECTROSTATIC CHARGE

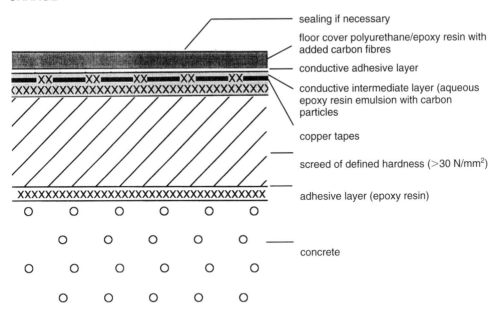

FIGURE 9.5. Electrical discharge plastic floor system.

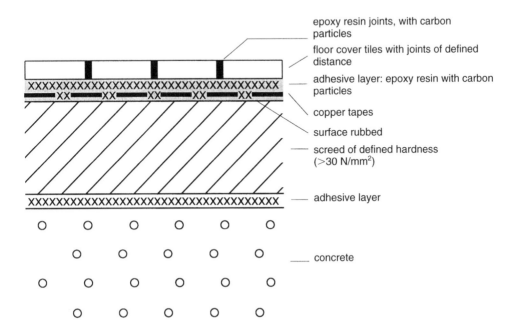

FIGURE 9.6. Electrical discharge tile floor system.

OTHER SUITABLE FLOOR SYSTEMS

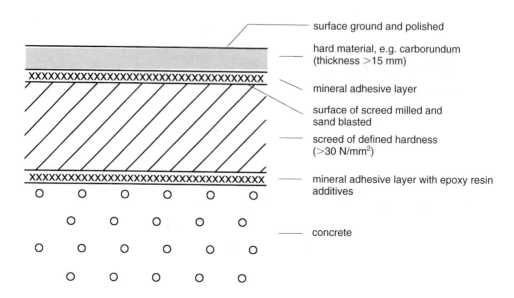

surface ground and polished

hard material, e.g. carborundum (thickness >15 mm)

mineral adhesive layer

surface of screed milled and sand blasted

screed of defined hardness (>30 N/mm^2)

mineral adhesive layer with epoxy resin additives

concrete

FIGURE 9.7. Electrical discharge concrete floor with a layer of hard material.

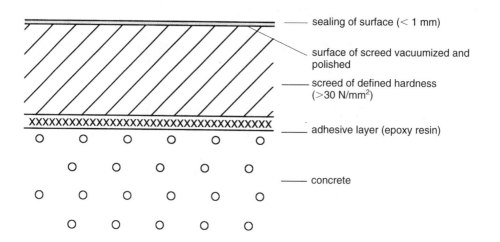

sealing of surface (< 1 mm)

surface of screed vacuumized and polished

screed of defined hardness (>30 N/mm^2)

adhesive layer (epoxy resin)

concrete

FIGURE 9.8. Plastic sealed concrete screed.

thickness to remain upstanding when lightly held. They are usually joined together either by an H-section of suitable material such as anodized aluminium or plastic, or interlocked together by various methods. These walls are typically 50 mm in thickness having an inner core sandwiched between outer cleanroom compatible surfaces. Typical examples of inner cores are as follows: gypsum plasterboard, rock wool, glass wool, and polystyrene foam, although other suitable materials can be used. Outer cleanroom compatible surfaces are materials such as mild steel covered with epoxy paint or powder coated, aluminium, stainless steel, and plastic sheet, which is laminated or formed round the core materials.

Figures 9.9–9.17 show typical cleanroom wall elements and their constructive details such as core materials, joints, connections between wall elements, connections to the ceiling and floors, etc. High standard sandwich elements are shown in Figures 9.9–9.12. These types of elements are manufactured from high-quality materials to very precise tolerances. Joint widths between the elements are less than 0.1 mm and there are therefore no visible silicon joints. Very smooth surfaces are presented to the cleanroom. Illustrated in Figures 9.13–9.17 are elements of a good standard of construction. These elements use less expensive materials and are not so precisely sized so that silicon joints can be seen.

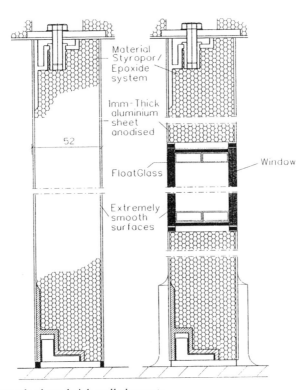

FIGURE 9.9. High standard sandwich wall elements,.

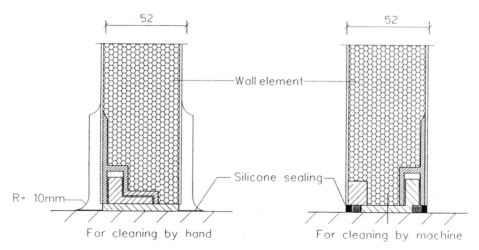

FIGURE 9.10. Detail of high standard sandwich elements showing two alternative floor connections suitable for machine and hand washing.

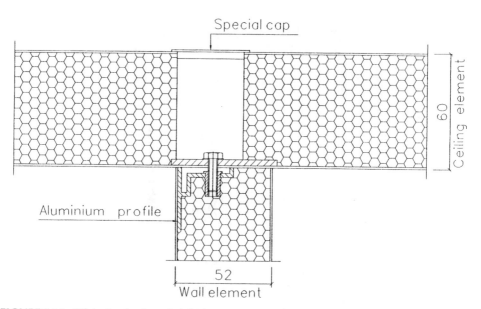

FIGURE 9.11. High standard sandwich element: connection between ceiling and wall.

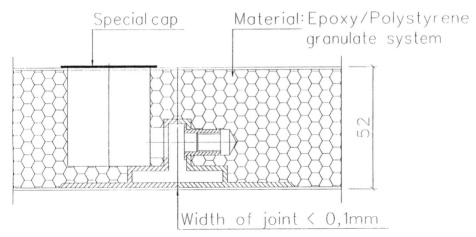

FIGURE 9.12. High standard sandwich element: connection between wall elements.

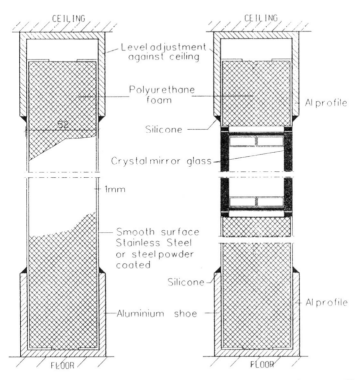

FIGURE 9.13. Good standard sandwich elements with proper connections to ceiling and floor: suitable for wet production *and machine cleaning*.

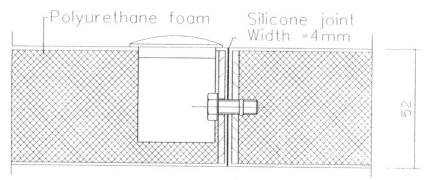

FIGURE 9.14. Good standard sandwich element: connection between wall elements.

Door Systems

Door systems in cleanrooms should have as few horizontal surfaces as possible. They should have abrasion-proof mechanisms, no horizontal surfaces, no abrasions on the floor, and as few door fitments, such as door handles and locks, as possible. The entire surface should be smooth, easy to clean, as well as resistant to cleaning agents and disinfectants.

Ceiling Systems

The *in situ* construction of cleanroom ceilings is quite unusual. Cleanroom ceilings are normally constructed using prefabricated methods. For high and medium cleanroom standards two different ceiling systems can be considered. These are as follows:

1. A system consisting of panels in which a sheet metal sandwich encloses an inner core of suitable material. An example of such a system is shown in Figure 9.18. This shows a uniform aluminium/polystyrene-epoxy sandwich system with a typical thickness of about 65 mm and a bearing capacity of about 1500 N/m^2. Shown in Figure 9.19 is a drawing of a lighting element used in such a system. These panels are interlocked together in a gas-tight manner without joints. All electrical and pneumatic services can easily be integrated into the sandwich elements.
2. Grid ceiling systems consisting of a light metal framework with fastenings for lighting installation, blank panels and filter frames as well as openings for cables and sprinklers etc. The points of intersection are sealed to prevent the passage of air and particles and the filter frames are installed tightly within their housings. The principle components of such a system are shown in Figure 9.20.

Shown in Figures 9.21 and 9.22 are two examples of a gas-tight grid ceiling system.

Another variation of the grid ceiling system is often used in semiconductor cleanrooms. In this system the grid has a channel with a gel in it. The filter and blank panels have knife edges which sit into the gel, thus preventing the leakage that can sometimes occur through poorly seated gaskets. This type of system is discussed in Chapter 8 of this book and illustrated in Figures 8.16, 8.17 and 8.18.

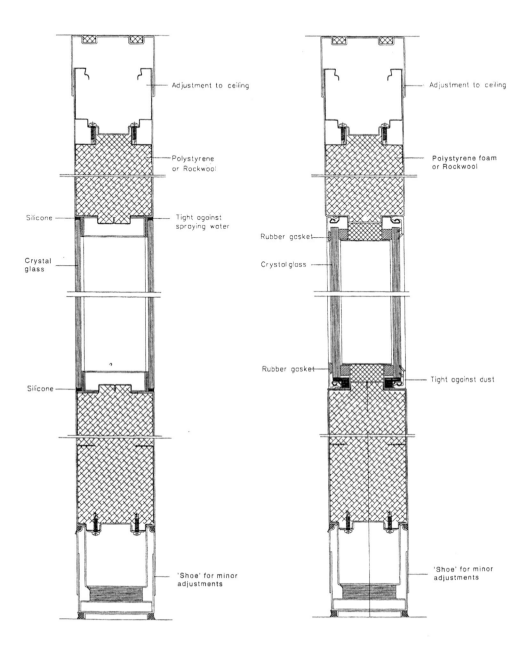

FIGURE 9.15. Good standard sandwich wall element suitable for a *wet* environment and machine cleaning.

FIGURE 9.16. Good standard sandwich wall element suitable for a *dry* environment and machine cleaning.

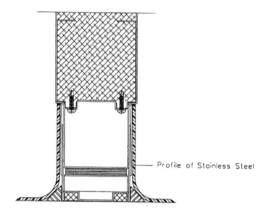

FIGURE 9.17. Good standard sandwich wall element suitable for hand cleaning.

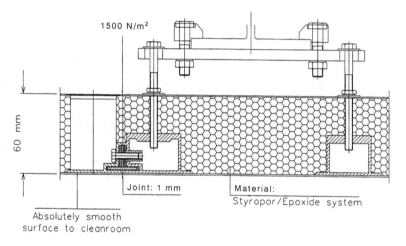

FIGURE 9.18. High standard sandwich ceiling element with a stable suspension to roof.

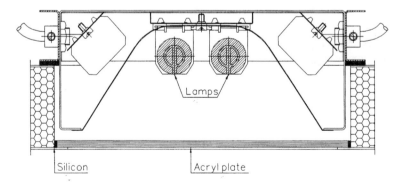

FIGURE 9.19. Detail of high standard lighting element.

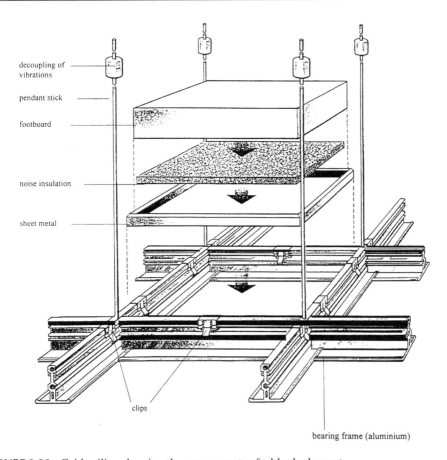

decoupling of vibrations

pendant stick

footboard

noise insulation

sheet metal

clips

bearing frame (aluminium)

FIGURE 9.20. Grid ceiling showing the components of a blank element.

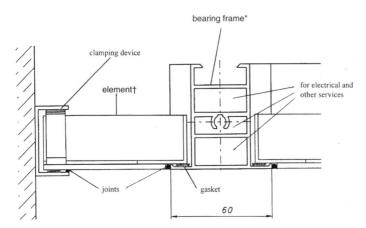

bearing frame*

clamping device

element†

for electrical and other services

joints

gasket

60

FIGURE 9.21. Grid ceiling: connection to wall and position of elements. *Bearing frame with a section profile suitable for a gas-tight installation of electrical pneumatic services. †The element can be a filter, a blank, lighting fixture, etc.

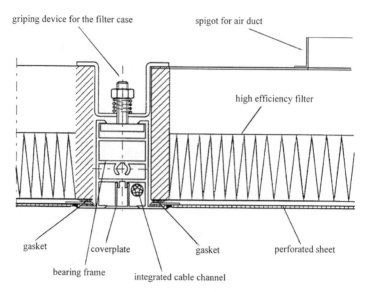

FIGURE 9.22. Grid ceiling: fixing of high efficiency filters.

For cleanrooms of a *lower* standard, grid ceiling elements on a supporting framework without tightening devices (in this case the lamps are suspended or integrated) is sufficient. In some cases a suspended panel system is sufficient. However, there is no possibility of maintenance and repair above the ceiling, i.e. cleanroom production must be stopped during these activities.

POSSIBLE CLEANROOM MATERIALS AND CONSTRUCTIONS

Table 9.1 shows some examples of possible cleanroom construction components that could be used with respect to different standards of cleanroom. Table 9.2 presents possible components of cleanrooms with good properties for minimizing static charge. The complete construction should be checked in the as-built state for its effectiveness against static charge. For areas with high cleanliness requirements, where the anti-static requirements cannot be met for the surface finish, an air ionization system can be provided.

MEASURES TO BE TAKEN DURING THE CONSTRUCTION AND ASSEMBLY OF CONSTRUCTION

The materials and components used in the construction of a cleanroom should not be delivered to the construction site in a dirty state. As well as not being externally dirty, dirt should not be 'built in' during manufacture, since over the lifetime of the cleanroom this contamination could enter the cleanroom. All components and materials used in the construction of the cleanroom should therefore be manufactured, packed, transported, stored and inspected, before use, to ensure their cleanliness.

It is also necessary, during construction, that measures are taken to ensure that the contamination generated in the course of assembly and construction work is contained

TABLE 9.1. Examples of possible cleanroom construction components.

High requirements

Materials for surfaces of wall and ceiling systems
- stainless steel, polished, ground or blasted
- aluminium, anodized (>20 μm thickness) or enamelled
- sheet steel, enamelled or powder coated
- conductive plastic panels

Ceiling systems / construction features
- light metal framework with integrated fastenings for the lighting installation and others, with openings for cables, sprinklers, etc. airtight at points of intersection
- panel ceiling on supporting framework, airtight, with integrated lighting installation and others, with non-abrasive sound absorbent layer if necessary.

Floor coverings on screed (with a defined adhesive strength between screed and layer)
- non-shrinking plastic sheets with hot welded joints
- layer of ceramics, joints of plastic material
- non-shrinking coatings of resins (e.g. epoxy resin, polyester or polyurethane), surface-sealed

Raised floors
- Coverings of non-shrinking plastic on plates standing above ground of sealed concrete

Medium requirements

Materials for surfaces of wall systems
- sheet metal, powder-coated
- zinc plated sheet metal, enamelled or powder-coated
- plastic panels
- plastic coated walls

Ceiling systems / construction features
- framework with fastenings for the lighting installation, with openings for cables, sprinklers, etc. Filter cases inserted from the bottom, or fastened against their gaskets.

Floor systems (defined adhesive strength between screed and layer)
- layer of hot welded plastic sheets or elastic floor tiles, glued on screed
- tiled flooring placed on screed or on concrete
- screed of concrete with an integrated layer of hard materials (e.g. carborundum) ground and polished
- concrete screed with plastic coating.

Raised floors
- same as high requirements

Low requirements

Materials for surfaces of wall systems
- gypsum board coated
- same as for median requirements

Floors
- layer of hot welded plastic sheets on screed
- concrete screed with a plastic coating
- concrete screed with a layer of hard material

TABLE 9.2. Components for minimizing static charge.

Floors

- Plastic floor coverings with the required electrical characteristics on raised metal floor constructions, grounded via resistors, if necessary
- Conductive resin coverings (e.g. epoxy resin, polyurethane, polymethylmetacrylate): multi-layer structure with added carbon fibres, carbon or aluminium particles
- Conductive tiled flooring

Example for semiconductor industry (layers are given from the base to the surface)
- coating on ground
- mounting of copper tapes in polyurethane conductive layer with additives (e.g. carbon particle dispersion)
- polyester resin surface layer with aluminium powdered grit
- electric welding of the aluminium additives
- connection of the copper tapes to the grounding frame with the possibility of installing resistors for adjustment

Example for pharmaceutical industry (layers from base to surface)
- coating on ground with imbedded copper lattice or tapes
- conductive thick bed plaster (carbon particle dispersion)
- tiled flooring on conductive screed (tiles with the necessary content of spinel or tiles with a glazing of spinel on the top and at the side)

Walls

- Metal, grounded via resistors, if necessary
- Plastic panels with conductive coating.

Fittings

- Tables with conductive plastic surface and conductive frame or metal tables
- Machine surfaces
- Metal, grounded via resistors, if necessary
- Plastic sheets with conductive coating.

and removed, so as to limit undue contamination of surrounding areas. Appropriate means of containment may include the use of temporary screens and walls, and pressurization of critical zones, with provisional use of temporary 'sacrificial' filters in the air handling system(s). Such filters, installed to protect the clean environment and air handling systems from outside contaminants, and to permit their initial pressurization and operation, are intended to be removed and replaced by filters of the appropriate grade at the agreed stage or stages of start-up, before construction approval and subsequent operational use of the installation.

Continual or frequent cleaning during construction and assembly should be planned, undertaken and controlled, with the aim of preventing undue build-up of contaminants in any part of the installation, and so facilitating the essential final cleaning before start-up.

BIBLIOGRAPHY

VDI guideline 2083/part 4: Cleanroom technology, Surface cleanliness (2/96). GAA RR in DIN and VDI; Beuth Verlag GmbH, Berlin.

Working papers by members of the German Mirror Committee of ISO TC 209/WG 4. Authors: G. Dittel, c/o Ing. Büro Dittel; H. Eidam, c/o Turbo Lufttechnik; A. Machmüller, c/o Ineliol and E. Sirch, c/o Bayer AG.

10 Purification Techniques for Clean Water

T. HODGKIESS

INTRODUCTION

Cleanroom users utilize process water which is usually supplied from a public water system but can be of widely varying quality depending upon the source of the water. In general, water derived from a deep well or borehole will contain a relatively high loading of dissolved salts due to leaching of minerals through which the water has percolated. These salts (often termed 'minerals') comprise charged ions and their total content is essentially what is commonly termed the total dissolved solids (TDS) content or a water. Such well water will usually contain relatively small amounts of dissolved organics, suspended solids, and biological organisms as a result of a natural filtering action that it has undergone. In contrast, surface water (i.e. moorland catchment or river) will be much lower in dissolved salts but will have relatively high burdens of suspended solids, dissolved organics (with seasonal variations), and bio-organisms. For illustration purposes, some analysis details of public waters from two different areas are presented in Table 10.1.

Specifications for final process water for various cleanroom purposes are dealt with in detail in a later chapter but, as a means of initial comparison with the types of water supply exemplified above, the following figures are provided at this point to provide an indication of the levels of water purity required for final rinse water for micro-electronics:

- TDS limit specified in terms of a water resistivity > 18 megohm-cm at $25°C$ (equivalent to a sodium chloride content of < 0.004 mg/l),
- TOC < 30 microgram/1,
- Bacterial count < 1 CFU/ml,
- Particle count 0 (1.0–2.0 μm), < 50 (0.5–1.0 μm)

Although generally less stringent than for micro-electronics, high levels of purity are also required for the pharmaceutical industry. This is shown in the following United States Pharmacopeia specification for 'purified water':

- Water resistivity > 0.8 megohm-cm (about 0.7 mg/l TDS),
- TOC < 500 μm/l,

Cleanroom Design. Edited by W. Whyte © 1999 John Wiley & Sons Ltd

- bacteria < 100 CFU/ml.
- with additional requirements relating to water for injection: bacteria < 0.1 CFU/ml and a pyrogen content of < 0.25 endotoxin units/ml.

It is thus clear that substantial treatment of the water supply will generally be necessary for all cleanroom purposes and that the detailed water-treatment system and layout will vary somewhat from plant to plant. Details of the design and operation of a typical plant are presented in a later chapter and it is the purpose of this chapter to survey the range of possible purification techniques which can form components of overall systems for the production of clean water. The range of undesirable constituents, and some of their removal processes considered in this chapter, is summarized in Table 10.2.

The listing immediately signals the potential for some techniques to eliminate more than one type of impurity. This is particularly the case for membrane separation processes. These comprise a number of techniques that mainly involve a range of 'filters' to facilitate the rejection of constituents of progressively decreasing size, as indicated in the approximate representation in Figure 10.1.

TABLE 10.1. Analysis of public water from two different areas.

Constituent	Supply 1	Supply 2
Total dissolved solids, 'TDS' (mg/l)	19	315
Chloride (mg/l)	7.0	54
Sulphate (mg/l)	3.9	88
Carbonate (mg/l)		33
Nitrate (mg/l)		38
Sodium (mg/l)	2.8	44
Potassium (mg/l)	0.3	10
Magnesium (mg/l)	0.8	15
Calcium (mg/l)	4.2	33
Silica (mg/l)	1.3	6
Residual chlorine (mg/l)	0.2	0.2
Total organic carbon 'TOC' (mg/l)	2.0	
Colour (Hazen units)	10	7
pH	9.2	7.4
Bacterial count (CFU/ml)		5

TABLE 10.2

Constituent:	Particulates	Dissolved salts (ions)	Dissolved organics	Bacteria
Possible removal techniques	Coarse filters Sand filters Membranes	Distillation Membranes (RO, ED) Ion exchange	Act. carbon Membranes Ion exchange	Membranes UV radiation Chemicals Heating

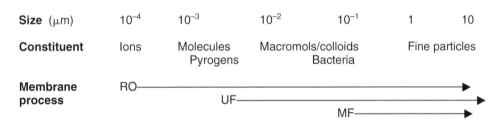

Size (μm)	10^{-4}	10^{-3}	10^{-2}	10^{-1}	1	10
Constituent	Ions	Molecules Pyrogens	Macromols/colloids Bacteria		Fine particles	

FIGURE 10.1. Approximate representation of impurity sizes in relation to some membrane separation processes. RO = Reverse osmosis; UF = ultrafiltration; MF = Microfiltration.

The detailed description of water-purification techniques will commence with the removal of the dissolved ionic constituents ('salts') as a convenience since the processes involved tend to have additional potential for the elimination of other types of impurity and hence are mentioned again in later sections of the chapter.

REMOVAL OF DISSOLVED IONS

Processes that remove the ionic constituents of water are alternatively called desalting, desalination, or demineralization, and it is possible to achieve this type of purification either by removing the pure water from the salt solution or by removing the salt from the solution. Examples of methods involving the removal of water from the salt solution are:

- *Distillation*, which provides, under ideal conditions, automatic separation of the pure vapour leaving the salt in the unevaporated solution.
- *Freezing*, which produces ice crystals which themselves do not contain any salt but have to be physically removed from the saline solution and washed.
- *Hydration*, which involves the formation of solid hydrates which after removal and washing can be decomposed to yield pure water.
- *Reverse osmosis*, which involves forcing pure water molecules, under the influence of an applied pressure exceeding the osmotic pressure of the solution, through an appropriate semi-permeable membrane.

Methods involving the removal of salt from the solution are:

- *Ion exchange*, which involves passing water through a column containing tiny resin particles composed of a substance containing (typically) weakly bound hydrogen and/or hydroxyl ions which can exchange for impurity ions in the water.
- *Electrodialysis*, which involves applying an electric potential across the water thus causing the oppositely charged ions to move in opposite directions. The presence of an alternating array of membranes which will allow the preferential passage of either positively charged ions (cations) or negatively charged ions (anions) results in the water in alternate channels between the membranes becoming purified.
- *Piezodialysis*, which involves passage, under an applied pressure, of salt through membranes permeable to both cations and anions but impermeable to water.

Of the above processes, freezing, hydration and piezodialysis have never become established on a commercial scale. Until the early half of the 20th century, distillation was the only practical means used to produce good quality water, but by the late 1930s ion exchange had become established followed by electrodialysis in the 1940s and 1950s and eventually by reverse osmosis in the 1960s.

Distillation is still widely used in the pharmaceutical industry partly on account of its perceived greater certainty in removing bacteria, viruses and pyrogens from water. However, in the microelectronics industry, the two main desalination processes utilized are reverse osmosis and ion exchange, with electrodialysis receiving consideration in some circumstances. These four processes will be described in more detail in the following sections.

Distillation

The basic operation of purifying saline water, by boiling it and collecting the condensing vapour to yield a product theoretically free from non-volatile impurities, has been practised for centuries. The process has the added feature of providing a sterilized product and can be used to treat a wide range of feedwaters. However, some carryover of dissolved and colloidal impurities into the product water does occur and the process also has certain other disadvantages, such as an inability to remove volatile organics and an inherently high energy consumption (largely associated with the high latent heat of evaporation of water). Indeed, the evolution of the design of commercial distillation plants has been driven by the need to reduce the energy consumption. One obvious way of doing this (Figure 10.2) is to use some of the condenser cooling water as feed for the evaporator. Simple distillers of this type, but usually electrically heated by an immersion heater, are widely used for small-scale production in general laboratories.

Further improvements in performance of a distiller are achieved by employing a succession of evaporation/condensation steps in one plant thus effectively re-using the energy input associated with one batch of heating steam. There are basically two ways of achieving this aim, i.e. by using Multiple-Effect Boiling or Multistage Flash Evaporation. These two processes, together with a further energy-reducing process, Vapour Compression, are described below.

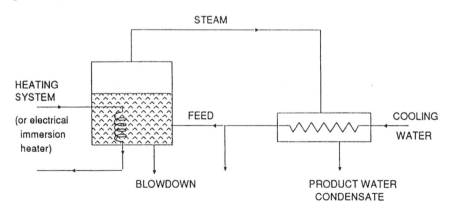

FIGURE 10.2. Flow diagram of a simple distillation unit.

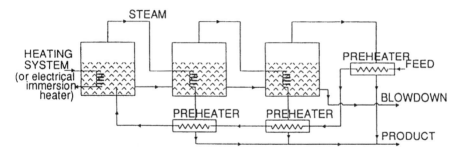

FIGURE 10.3. Flow diagram of three-effect ME evaporator.

Multiple-Effect (ME) Evaporation   In a ME distiller, steam, generated at a relatively high pressure (perhaps several bar) from the feedwater in the first effect, is used to boil the saline water at a lower pressure and temperature in the second effect and so on. The condensate from the steam-side of all effects (except the first) represents the purified product. A three-stage ME evaporator is shown schematically in Figure 10.3.

The thermal performance of a ME evaporator, is directly proportional to the number of effects. Consequently, the choice of the number of effects represents an optimization between increased capital cost and reduced operational energy costs as the number of effects increases. There are various designs of commercial ME distillers. They often comprise vertical distillation columns arranged with anything from three to eight separate effects with the first effect being heated either by an external primary steam source or by utilizing an electrical immersion heater.

Improvements in performance of ME units can be obtained by arranging for the evaporation to occur from thin films of the feedwater on either the internal or external surfaces of tubes. Thus, in the horizontal falling-film evaporator (Figure 10.4), the

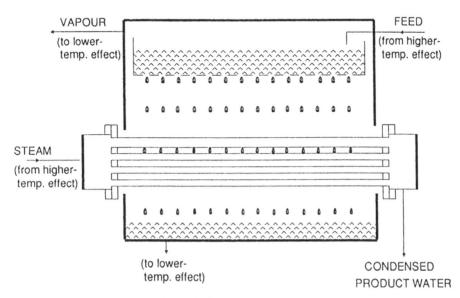

FIGURE 10.4. One effect in a horizontal falling-film evaporator.

preheated feed is sprayed over the outside of horizontal tubes and is partly evaporated by steam (supplied from a higher-temperature effect) condensing inside the tubes. In another variant, the vertical-tube evaporator, evaporation occurs from a falling film of saline water on the inside of vertical tubes with the steam being supplied to the outside of the tubes.

Multistage Flash Distillation (MSF)   In this type of plant the feedwater is pumped into a heater at such a pressure that it does not boil, before passing through an orifice into a large chamber. The reduction in pressure accompanying this process causes a fraction of the water to flash off as steam. The saline water then passes, via another orifice, into an adjacent 'flash chamber', at a yet lower pressure, where more flashing occurs and so on down the plant. Although a dominant process for the large-scale production of potable water from seawater in various parts of the world (especially the Middle East), MSF is not usually favoured for the small-plant applications typical of pharmaceutical manufacturers.

Vapour Compression (VC)   Another approach to reducing the energy consumption in a distiller is to compress (rather than condense) the vapour produced from a conventional distillation stage. The effect of compressing the vapour is to heat it to a temperature higher than the boiling point of the feedwater from which it was originally evaporated. Consequently, the compressed vapour can be returned to the same distillation chamber in which it originated and used in place of the external, primary heating steam in order to evaporate an equivalent amount of new steam from the feedwater as it, itself, is condensing. The cycle is continuously repeated. The process is shown schematically in Figure 10.5 which indicates that both the distillate product and the waste brine are used to preheat the feedwater in a heat exchanger. Once started, the VC process consumes relatively little energy—mainly the power required to operate the compressor.

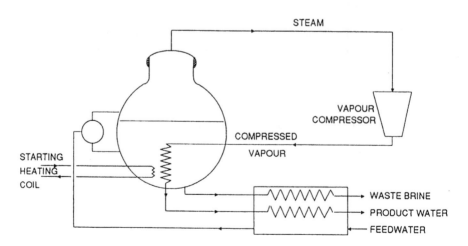

FIGURE 10.5. Schematic representation of a vapour-compression evaporator.

There are two distinct types of VC plant depending on the type of compressor:

1. *Mechanical vapour compression.* This utilizes a standard compressor such as a centrifugal compressor or a Roots blower. This approach is favoured when electrical energy is available.
2. *Thermal vapour compression.* Here the successive expansion and contraction of steam flowing through a system of nozzles results in the required increase in vapour pressure. This variant has the advantage of requiring fewer moving parts and is favoured on sites where a supply of medium-pressure steam is available.

Commercial VC units may be single effect or multiple effect in which the vapour from the lowest-temperature effect is compressed and returned as heating steam to the first effect.

Reverse Osmosis

Basic Principle If a solution of a non-volatile solute (such as sodium chloride) is separated either from a sample of pure solvent (such as water) or from a more dilute solution by a membrane which is permeable to the solvent but impermeable to the solute, there will be a tendency for the spontaneous passage of solvent through the membrane into the more concentrated solution under a driving force known as the *osmotic pressure* (Figure 10.6(a)). This transport of water molecules, diluting an aqueous solution, will continue until the excess pressure thus produced by this inflow on the concentrated solution side has built up to a magnitude equal to the osmotic pressure at which point the flow ceases (Figure 10.6(b)) due to the establishment of thermodynamic equilibrium across the membrane.

Desalination can be accomplished by reversing the naturally-occurring process shown in Figure 10.6(a). Thus, if a pressure greater than the osmotic pressure is applied to the concentrated solution side, then pure solvent will pass through the membrane producing a purer water product, Figure 10.6(c). This is the basis of reverse osmosis (RO) and a simplified representation of a RO system is shown in Figure 10.7.

The value of the osmotic pressure of any solution is directly proportional to the concentration of the solute in the solvent, that is, to the TDS of an aqueous solution.

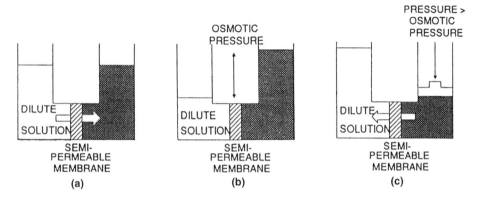

FIGURE 10.6. Osmosis and reverse osmosis.

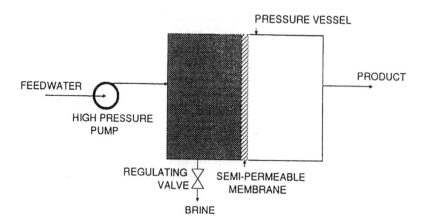

FIGURE 10.7. Simplified representation of reverse osmosis system.

Thus, the minimum required pressures (and hence the energy costs) for reverse osmosis increase with the salinity of the water being treated. As examples, the osmotic pressures of solutions of 100 mg/l, 0.8% and 3.2% sodium chloride are about 0.08, 6.8 and 27 bar, respectively. Whilst the osmotic pressure represents the minimum required pressure to carry out reverse osmosis, commercial units require operational pressures significantly in excess of this because the water flux obtained is directly proportional to the magnitude by which the operational pressure exceeds the osmotic pressure:

$$\text{water flux (in, say, m}^3/\text{m}^2.\text{day}) = k_1 \, (\Delta\rho - \Delta\pi), \tag{1}$$

where $\Delta\rho$ and $\Delta\pi$ are the operational-pressure and the osmotic-pressure differentials, respectively, across the membrane and k_1 is the pure-water permeability of the particular membrane utilized.

Another important requirement of a RO unit is the attainment of a high salt rejection by minimizing the salt flux across the membrane.

$$\text{salt flux} = k_2 \, \Delta C, \tag{2}$$

where k_2 is the salt permeability constant and ΔC is the difference in salt concentration in the feed and the product waters. The salt rejection is basically defined as follows:

$$\text{salt rejection} = (C_f - C_d)/C_f,$$

where C_f is the salt concentration in the feed and C_d is the salt concentration in the product. In the above relationship the salt concentrations are usually expressed in terms of the total salt content, TDS, but it should be noted that RO membranes possess somewhat varying abilities to reject different ionic species. In general, superior rejections are obtained for multivalent ions than for univalent ions.

Notice, incidentally, that an indirect consequence of achieving a higher H_2O flux is to increase the salt rejection. Hence, increasing the operational pressure yields two benefits, namely increased productivity and a purer product.

Another important operational factor is the 'recovery' which is defined as:

$$\text{recovery} = (\text{product flow}) / (\text{feed flow})$$

and is obviously a measure of the proportion of product water extracted from the feed. It is clearly desirable to maximize the recovery, which is often around 70%–80%. The upper limits to achievable recovery are dictated by the TDS of the feed (increases in which tend to reduce obtainable recovery) and by the detailed ionic composition of the feed (which, for example, can limit recovery on account of a higher susceptibility to fouling of the RO membrane by precipitation of low-solubility salts such as calcium carbonate).

Another relevant operational parameter is the temperature. Increases in temperature tend to reduce salt rejection but, more importantly, facilitate increased flux by reducing the water viscosity and hence promoting enhanced fluid flow through the membrane.

Reverse-Osmosis Membranes   The initial commercial development and subsequent advancements in reverse osmosis can be directly linked to the availability of membrane materials in appropriate forms. Semi-permeable membranes, suitable for commercial water purification by RO, consist of thin films of polymeric materials which first became available around 1960. The aim, for desalination purposes, is to produce a membrane which is highly permeable to water whilst being highly impermeable to salts and the basic difficulty is that attempts to close the paths for salt generally tend also to cause increased resistance to water flow. The breakthrough in membrane technology came when methods were found to produce a cellulose acetate membrane which combined the two required properties by having a very thin dense skin (the 'active layer') on top of a porous substrate. The thin skin (of this so-called 'asymmetric membrane') had good salt-rejecting properties and, being only around 1 μm thick, also had reasonable water flux—the porous substrate providing mechanical support for the thin layer without hindering the passage of water.

The asymmetric cellulose acetate (CA) membranes typically exhibit salt rejections between 92% and 97% and water fluxes in the range 0.4–0.8 m^3/m^2 per day. More recent membrane development has led to improved fluxes and separation characteristics and has also been aimed at utilizing materials which can be used over a wider pH range and have lower tendencies towards compaction of the porous layer (which causes reduced flux) than the CA membranes. Of the range of materials which have been studied and developed, the most successful alternatives to cellulose acetate have been *polyamides*. However, it is important to recognize that polyamide membranes are vulnerable to irreversible damage when exposed, for even short periods, to water containing chlorine. Thus, feedwater must be effectively dechlorinated prior to supply to a RO module utilizing polyamide membranes.

Many of the recently developed reverse-osmosis membranes have been produced as thin-film 'composite membranes'. The essential difference between such a membrane and a conventional sheet CA membrane is that, whereas the latter contains an active and support layer made in one series of operations, in composite membranes these two layers are made in separate stages and often consist of different polymers (Figure 10.8). Hence it is much easier to secure improvements in the performance of each layer

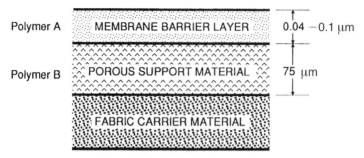

FIGURE 10.8. Schematic representation of a composite membrane.

without the risk of reducing the performance of the other layer. One particularly useful feature of composite-membrane construction is the ability to produce much thinner active payers than is possible with the asymmetric form and the overall effect is the availability nowadays of composite membranes with better salt rejection without any reduction in the water flux compared to that obtained with cellulose acetate. Different versions of such thin-film composite membranes are manufactured for desalination of seawater or (more relevant to this book) for treatment of brackish water of say a few hundred to a few thousand mg/1 TDS. For example such brackish water RO membranes operating at perhaps 8 bar pressure achieve rejections of 98%–99% for most salts.

Fouling of RO Membranes   As indicated in Figure 10.1, a RO plant can be considered almost to be an ultimate filter for impurities in a water. Whilst this feature is, on the one hand, an attribute of the process, on the other hand it represents a weakness. This is because a RO membrane is prone to becoming fouled with a number of feed constituents. Despite the sweeping action of the feed flowing through and along the membrane, a feed containing a high burden of suspended matter, biological organisms, or dissolved salts near to their solubility limits will result in the settlement of these constituents on the membrane surface. This fouling layer drastically degrades the performance of the membrane in ways that are 'signalled' by a progressive increase in transmembrane pressure, increase in product TDS and decrease in product flux to an extent that frequent intermittent cleaning operations will at least be necessary and, in extreme cases, complete membrane replacement will be required.

Salt Concentration Effects During RO Operation   Two consequences of the action of a RO membrane in allowing preferential passage through it of H_2O molecules from a saline solution, are:

1. As water flows along a RO membrane, it is progressively increasing in salt content in the bulk stream.
2. Additionally, there is a tendency for the rejected salt to accumulate in the concentrate boundary layer adjacent to the membrane surface. This phenomenon is known as *concentration polarization*.

Both these phenomena cause a reduction in the driving pressure, $(\Delta\rho - \Delta\pi)$, equation

(1), and an increase in the salt flux, see equation (2), i.e. they contribute to a lowering of the average performance of the unit compared with the performance calculated on the basis of the feed composition. An important feature of a detailed RO plant design is to minimize the extent of concentration polarization.

Reverse-Osmosis Plant Configurations   Reverse-osmosis plants are normally built on a modular basis. The membranes are commercially available in relatively small, modular units consisting of the membrane mounted on a supporting system within a pressure vessel complete with feed, product and concentrate connections. The design and engineering of reverse-osmosis systems falls into two, often quite separate, parts, i.e. (i) the design and manufacture of membranes and membrane modules (this activity involving a small number of large companies) and (ii) the design and construction of complete reverse-osmosis systems making use of the commercially available membrane modules (an activity involving a far greater number of organizations).

The objectives of membrane module design include:

- The provision of a large membrane area in a relatively small module volume.
- The provision of sufficient membrane support to withstand the large pressure differences across the membrane.
- The minimization of the effects of polarization.
- The minimization of frictional pressure losses in both concentrate and product channels.

These requirements have led to the evolution of four basic module designs: plate and frame, tubular, spiral wound and hollow fibre of which the latter two are predominant in water purification applications and hence will be described below.

Spiral Wound Systems   This type utilizes membranes in a convenient flat form. The membrane is cast onto a fabric support and then two of these fabric-supported membranes are glued together with a porous material between them. This porous material supplies the subsequent route for the product water after its passage through the membrane. The resulting 'sandwich' is then glued together on three of its edges with its fourth edge being glued to a central tube which, containing appropriately spaced holes in its wall, acts as a collector for the product water. A flexible spacing mesh is then laid on top of the sealed membrane sandwich and the whole lot rolled up around the product collection pipe to form a spiral wound membrane unit (Figure 10.9). The unit is then mounted in a cylindrical pressure vessel usually constructed from glass-fibre reinforced polymer. The feed flows axially along the spacing mesh and the permeate via the porous backing material to the central collector. All kinds of membranes have been used in the spiral wound configuration—cellulose acetate, polyamide and other composite types.

Hollow Fibre Systems   These, as shown in Figure 10.10, utilize large numbers of fine tubes with diameters about the same as that of a human hair (the hollow fibres having outside diameters of about 80–200 μm and bores about a half of the outside diameter). This approach enables high membrane packing densities to be attained and has the

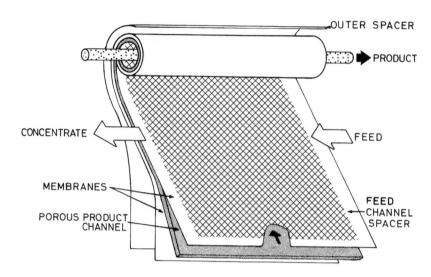

FIGURE 10.9. Spiral wound module.

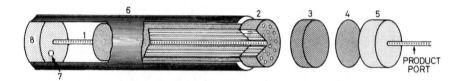

FIGURE 10.10. Hollow-fibre RO module (explanation of components given in text).

added advantage that, at such small diameters, the hoop stresses incurred by the high operating pressures are small enough for the membrane material to withstand without any mechanical support (provided the high pressure is on the outside of the fibre and the hoop stress is compressive). The membrane units, which are commonly called fibre bundles, are made up by winding a continuous fibre around a piece of backing cloth that is as wide as the fibre bundle is to be long. The backing cloth is then rolled up around a feedwater distribution pipe (1) with the fibres parallel to the pipe axis and the two ends of the fibre bundle so formed are set into epoxy-resin blocks. One of these blocks (2) is then cut away to expose the open ends of the fibres to allow the product water to emerge via the bores of the hollow fibres. This open-end tube sheet (2) is supported by a porous block (3) and a spacer (4) against the product end plate of the vessel (5). At the other end of the pressure vessel (6), into which the fibre bundle is located, the feed is introduced along the central pipe (1) and the concentrate is extracted through an off-centre port (7) in the end cap (8). The high-pressure feed is made to flow radially outwards across the fibres in the bundle. The permeate then passes through the fibre walls into the bores and flows along the bores and out at the open end.

The fluxes obtained (as low as 0.01 m^3/m^2/day) using hollow fibre membranes are two orders of magnitude lower than those of spiral wound systems. Also the operating pressures have to be kept lower because of the unsupported nature of the hollow fibre membranes. The low fluxes, however, do mean that polarization is not significant and are compensated for by the very high packing densities (around 30 000 m^2/m^3) obtained when using hollow fibres; membrane packing densities in spiral wound systems are of the order of 600 m^2/m^3. An important disadvantage with the hollow fibre configuration is that the concentrate passages through the fibre bundle are very narrow and may very easily become blocked by suspended solids in the feed. It is therefore necessary to have very fine filtration in the RO pretreatment system.

Module Arrangements   In either spiral wound or hollow-fine fibre form, high production rates are obtained by supplying the pressurized feed to an appropriate number of membrane modules arranged in parallel. Brackish water membranes can often achieve a suitable degree of desalting in one pass of feed through a module. However, if, as is often the case, a high plant recovery is important, then 'brine staging' is used. This involves using the reject brine stream from the first set of parallel modules as feed to a second set of parallel modules and so on.

Energy Recovery in Reverse-Osmosis Units   The main component of the energy requirement for a RO plant is in the high-pressure pump for the feedwater and an example of a way of saving energy is to utilize the power available in the concentrate flowstream to drive a turbine which, in turn, drives the main high pressure pump through a common shaft. Savings of around 30% on pumping costs have been claimed for seawater-RO units but for RO plant operating on relatively low TDS feed, the use of energy recovery is less attractive on account of the much longer pay-back time on the capital cost of the turbine etc.

Nanofiltration   This can be considered as a variation on the basic RO process. Nanofiltration utilizes membranes specially tailored to achieve high rejections of multivalent ions compared with univalent ones. Acceptable fluxes can be obtained at relatively low operational pressures and the process is sometimes referred to as 'loose RO'. Such rejection capabilities mean that a primary application of nanofiltration is for water-softening purposes involving the removal of ions such as calcium, magnesium, barium and strontium.

Ion Exchange

Basic Principles   This method involves the removal from solution of one type of ion and its replacement by an equivalent quantity of another ion of the same charge. Thus there are two modes of ion exchange:

1. *cation exchange*, in which some or all of the cations in the water (calcium, sodium, etc.) are removed, and
2. *anion exchange*, in which some or all of the anions (chloride, sulphate, etc.) are removed.

Figure 10.11 shows a schematic diagram of one type of ion exchange system. The top compartment or column (the cation exchanger), contains a huge number of tiny (typically 0.5 mm diameter) resin particles each of which contains weakly bonded and

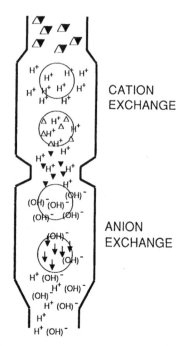

CATION
EXCHANGE

ANION
EXCHANGE

FIGURE 10.11. Cation and anion exchange columns in series.

thus replaceable H^+ ions and strongly bonded, fixed anions (not shown for the sake of clarity). As the feed water percolates through the column, cations of any species (shown as light triangles) in the water tend to be absorbed onto the resin particles in exchange for the weakly bonded hydrogen ions. So the water leaving the top column contains H^+ ions and the anions (shown as dark triangles) originally present in the water. This solution is now passed through the second, lower column (the anion exchanger) in which the resin particles contain weakly bonded, replaceable $(OH)^-$ ions and strongly bonded, fixed cations (again not shown). So, as the water percolates through the lower column, the anions in solution (including those derived from silica) exchange for the hydroxyl ions on the resin and the product water is purified containing essentially hydrogen ions, hydroxyl ions and water molecules, although, as described later, the process in practice is not absolutely 100% efficient.

The ion exchange reactions described above can be represented as follows:

Cation exchange $(H^+)_R + M^+ = (M^+)_R + H^+$ $\hspace{4cm}$ (3)

Anion exchange $((OH)^-)_R + X^- = (X^-)_R + (OH)^-$ $\hspace{3cm}$ (4)

where the subscript R represents the ionic species loosely bound in the resin particles, M^+ represents a singly-charged cation (such as Na^+) present in the water, and X^- similarly is a singly-charged anion (such as Cl^-) present in the water.

The ion exchange system shown in Figure 10.11, and described above, removes all types of impurity ionic species from a feedwater—a process often called 'demineralization'. Although this is the type of ion exchange which is most relevant to the production of high purity water (such as that required in electronic component manu-facture) and hence will receive most attention herein, it should be emphasized that ion exchange is often used in other applications where selective removal of certain ionic impurities is required. These will require a different ion exchange resin containing alternative replaceable ions to the hydrogen and hydroxyl ions shown in the set-up in Figure 10.11.

Resin Regeneration   It is clear that as water continues to pass through, for example, a cation exchange column, the number of hydrogen ions on the resin particles available for ion exchange declines with time. The resin becomes gradually exhausted, starting first at the entry to the column with the exhaustion zone gradually moving down the column. Eventually it is necessary to take the column out of service in order for it to be regenerated.

For a strong-acid (hydrogen-ion loaded) cation resin, regeneration can be carried out by passing through the column a strong acid such as hydrochloric acid which furnishes a high H^+-ion concentration in the water that causes reaction (3) to reverse (i.e. to proceed to the left) to eventually restore the resin to a condition suitable for further water purification.

Similarly, to regenerate a strong-anion $((OH)^-$-loaded) resin, a strong base such as caustic soda (NaOH) can be used.

It is clear that the frequency of regeneration will be very dependent upon the salt content (TDS) of the feedwater to the ion exchange columns. This factor establishes the effective role of ion exchange in the demineralization of water to that of final desalting of a feed that is already relatively low in ionic loading. To use ion exchange to desalinate a high-salinity water would be extremely uneconomic because of the neces-sary very frequent regeneration.

Ion Exchange Selectivity and Influence on Regeneration Practice   A particular cation-exchange resin does not possess equal tendency to remove all positively charged ions in a feedwater. Similarly, an anion-exchange resin has different affinities for different negatively charged ions. The number of charges on a particular ionic species (the 'valence' of the ion) is a major factor in determining the selectivity. In general, ion exchange resins loaded with hydrogen ions or hydroxyl ions have greater affinity for higher-valent ions than for lower-valent ones. In other words, such a cation resin would remove Al^{3+} selectively to Ca^{2+} and the latter selectively to Na^+. A more extensive comparison of the relative selectivities of both a typical strong-acid and a strong-base ion exchange resin is presented below:

$$Ba^{2+} > Ca^{2+} > Mg^{2+} >> (NH_4)^+ > K^+ > Na^+ > H^+$$

$$(SO_4)^{2-} >> (NO_3)^- > (HSO_4)^- > Cl^- > (HCO_3)^- > (HSiO_3)^- > (OH)^-$$

These selectivity considerations mean that 'slippage' or 'leakage' of impurity ions through an ion exchange column will preferentially involve the lower-valent ions because the more highly-attracted ions will tend to be stripped from the water first and hence be situated, during a service run, in the upper zones of a column. For example, with a feedwater containing Ca^{2+}, Mg^{2+} and Na^+, the cation exchanger, nearing the point of exhaustion, will tend to contain (Figure 10.12(a)) successive bands of Ca^{2+}, Mg^{2+}, Na^+ (and possibly H^+ at the very bottom). This situation causes problems in the subsequent regeneration operation, if the latter is carried out in the same direction as the flow in the exhaustion phase, because the Ca^{2+} ions stripped off the upper zone move downwards removing lower-affinity ions like Na^+ and this results (Figure 10.12(b)) in a remaining band of Ca^{2+} ions at the bottom of the bed which increase the leakage in the early parts of the next service run. Counterflow regeneration is a way of attacking this problem, by ensuring that the bottom of the bed is the most completely regenerated zone (Figure 10.12(c)). Indeed, counterflow regeneration confers several significant benefits. These are regeneration efficiency, savings in regenerant, working capacity and treated water quality enabling 'mixed-bed quality' (see later) to be almost attained with single-bed units. However, the regenerant practice with countercurrent regeneration is rather more complex. This is because the benefits of countercurrent regeneration depend on maintaining an undisturbed resin bed during successive service/regeneration cycles. There are basically two ways of achieving this.

1. One approach is to hold the bed down by the application of a flow of air or water

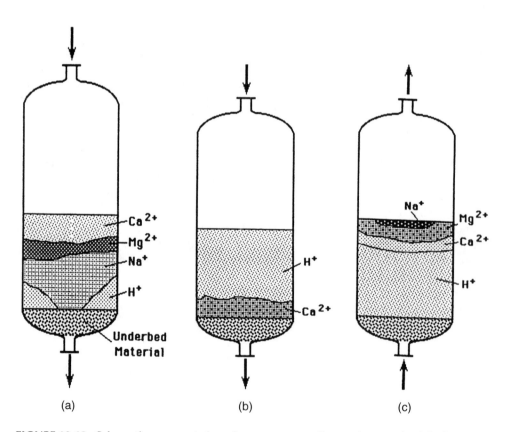

(a)   (b)   (c)

FIGURE 10.12. Schematic representation of consequences of ion exchange selectivity in a cation exchanger. (a) Towards end of service run; (b) after cocurrent regeneration; (c) after countercurrent regeneration.

from the top of the column during the regeneration phase depicted in Figure 10.12(c). A disadvantage of this method is the significant amount of so-called 'freeboard' space required at the top of the column.

2. The alternative strategy accomplishes a reduction in this wasted freeboard space (which tends to increase vessel size and required space for the water treatment system) by filling the vessel more completely and then raising the entire resin bed against an upper screen during upward regeneration. An inert resin layer is employed above the ion exchange resin in order to allow the passage of suspended solids and resin fines without loss of useful resin. Such a system is called a 'packed column' and, in some countercurrent designs, utilizes an upward service run and downward regeneration step.

Returning to the problem of leakage, a particular aspect of this in anion exchangers is that silica slippage can be troublesome because it is relatively weakly bound to the resin. Indeed, the monitoring of silica levels in the effluent of ion exchangers is sometimes used to signal the need for regeneration. Chloride slip can sometimes be associated with failure to remove organics upstream of the ion exchanger resulting in a decline in the performance of the anion resin due to fouling.

Degree of Deionization Attainable by Ion Exchange   Consider the operation of a cation-exchange column through which a feedwater is passing during a service run. Writing reaction (3) for a specific ion say Na^+

$$(H^+)_R + Na^+ = (Na^+)_R + H^+. \tag{5}$$

Reaction (5) proceeds to the right as Na^+ ions in the feed are exchanged for H^+ ions on the resin and this results in a gradual increase in concentrations of the species on the right of equation (5) and a decrease in the concentrations of the species on the left of the equation. As can be envisaged intuitively and deduced rigorously from chemical thermodynamic principles, these concentration changes progressively reduce the driving force for reaction (5) to proceed to the right until eventually the reaction comes to thermodynamic equilibrium and stops. A similar argument obviously applies to other cations (e.g. Ca^{2+}, Mg^{2+}) present in the feed. The consequence of this influence of thermodynamics is that there is a limit to the degree of deionization attainable and the water issuing from the exit of a cation-exchange column will still contain a residual concentration of impurity cations. Further elimination of these cations would require passing the partly purified water through another cation exchanger, and the conclusion reached is that *complete* elimination of the impurity cations from the water would require an infinite number of cation exchangers in series.

A similar argument holds for elimination of the impurity anions (such as Cl^-, $(SO_4)^{2-}$) with the minor modification that, for an anion exchanger arranged to receive the effluent from a cation exchanger, the anion exchange reactions generalized by (4) can proceed somewhat further to the right on account of further reaction of the so-produced $(OH)^-$ ions with the H^+ ions arriving from the upstream cation column:

$$(H^+) + (OH)^- = H_2O. \tag{6}$$

Indeed, this situation suggests a strategy for increasing the degree of deionization attainable in one ion exchange unit, i.e. to use a 'mixed-bed' column which contains both cation-exchange and anion-exchange resin. In such a unit, H^+ ions entering the percolating solution from the cation-exchange beads can be immediately consumed by $(OH)^-$ ions entering the solution from adjacent anion-exchange particles. Thus, reactions (3) and (4) can proceed much farther to the right, facilitating a much greater extent of demineralization from one mixed bed unit than from a single pair of separate cation-exchange and anion-exchange columns, hence reducing the amount of equipment necessary to achieve a certain desired level of purification. For this reason, the mixed-bed mode is preferred for the final stages of the production of ultra-high-purity water—although this benefit is gained at the expense of relative complexities in the regeneration cycle, as discussed in a later chapter.

ELECTRODIALYSIS

The Basic Process

Since dissolved salts exist as charged ions, the passage of a DC current through a saline solution will cause the positively-charged ions and the negatively-charged ions to migrate in opposite directions. If there is placed in the water a pair of membranes, one of which (a 'cation-permeable membrane') selectively allows the passage of cations and the other (an 'anion-permeable membrane') allows selective transport of anions, the water between the membranes can become desalinated.

An electrodialysis (ED) cell is shown in Figure 10.13. It contains electrodes at each end and a series of compartments or channels, of typically 1-mm widths, separated by membranes. Alternate membranes are anion permeable ('A' in Figure 10.13) and cation permeable ('C' in Figure 10.13). The membranes are thin sheets of polymer which have been treated to impart selective permeability and are in effect ion exchange resins in sheet form.

Under the influence of an applied DC potential between the electrodes, current flows within the ED cell, being carried by cations which tend to migrate towards the negatively charged electrode (cathode) and anions which tend to move in the direction of the positively charged electrode (anode).

To see how water purification can occur in such a cell, consider the smaller, sim-

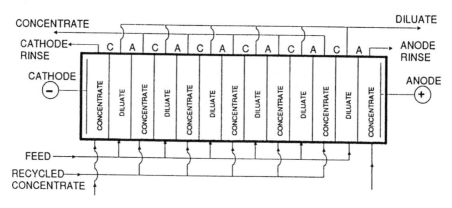

FIGURE 10.13. Electrodialysis process—general layout.

plified set-up shown in Figure 10.14 and, in particular, the events in compartment D_1. The various cations present in the water (Na^+, Ca^{2+}, etc.) can pass freely through the cation-permeable membrane at one end of the compartment and the anions can pass through the anion-permeable membrane at the opposite end. However, neither the cations nor anions can move out of adjacent compartments, F, because the membranes towards which they move (under the influence of the applied potential) are of the wrong type to allow passage of the ions. Nevertheless ions can escape from compartments D_2.

The mode of operation described above is sometimes called 'unidirectional ED' and units of this type are produced by some manufacturers. However, one of the major manufacturers of electrodialysis plant now uses a modification known as 'electrodialysis reversal' (EDR) which involves periodic reversal of the DC polarity about every 15 minutes or so. This modified form of operation reduces the tendency for membranes to become fouled by scales and slimes. Hence the need for chemical pre-treatment is reduced or eliminated but some extra valving is required on account of the switchover of diluate channels to concentrate channels and vice versa each time the polarity reverses.

Operational Factors   The rate of purification of water as it passes through the electrodialysis cell is proportional to the current flowing and also to the number of membrane pairs and is inversely proportional to the flow rate. For this reason, commercial electrodialysis stacks contain several hundred membrane pairs.

The energy requirements of the process are largely associated with the 'membrane

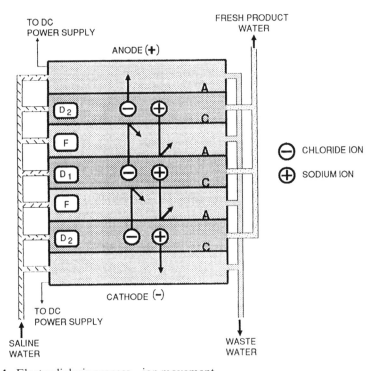

FIGURE 10.14. Electrodialysis process—ion movement.

potential', that is, the required voltage necessary to force ions across a membrane from the diluate to the concentrate against their *natural* tendency to traverse it in the opposite direction. The membrane potential is proportional to $\log(C_c C_d)$, where C_c and C_d are the salt concentrations in the concentrate and diluate, respectively. What this immediately tells us is that the greater the degree of purification required, the greater the membrane potential and hence the energy requirements for the process.

Moreover, a phenomenon known as 'concentration polarization' (with some general similarities to polarization in RO systems but having more important consequences on the diluate side of ED membranes) increases the energy requirements. Concentration polarization is especially prevalent when desalting at high rates, i.e. at high currents.

A further important aspect of the energy requirements is the voltage drop across the solutions in the electrodialysis cell and this becomes more appreciable the lower the salt concentration (i.e. the higher the resistance) in the diluate. Hence the production of highly purified water by this method *alone* would be hopelessly uneconomic and any role of electrodialysis in such applications would be prior to an ion exchange unit to reduce the feedwater TDS to a level at which ion exchange is more effectively utilized.

The Electro-deionization or Continuous Deionization Process

This can be thought of as modified electrodialysis or as a hybrid ED/ion exchange process. Its attraction is that it overcomes some of the important limitations/disadvantages associated with the separate ED and ion exchange processes.

As stated in the preceding section, the basic ED or EDR process is not very suitable for the production of water of low TDS—as required by the microelectronics and medical/pharmaceutical industries. A major reason for this is the high resistance of pure water which increases the energy consumption of the ED process. A way of surmounting this problem is to fill the diluate channels of the ED unit with a mixture of normal, mixed-bed, cation-exchange and anion-exchange resin beads. The basis of this approach is that, under the influence of the applied voltage, ions can move much faster through, and via, ion exchange particles than through the low-TDS water, i.e. the ion exchange beads provide a relatively high-conductivity path for ion transport in the diluate channels.

A schematic representation of such a continuous deionization (CDI) unit, in its simplest conceptual form, is shown Figure 10.15. In such a plant the concentrate is recycled in order to maximize both the conductivity and also the water recovery. Continuous deionization often represents a direct alternative to conventional ion-exchange columns for the production of low-TDS water because the former equipment tends to occupy less space than ion-exchange columns and can be operated for longer periods without the frequent regeneration of the resin that is necessary in ion-exchangers. This latter point relates to benefits associated with concern about handling of large quantities of acids/alkalis used for regeneration and problems with their disposal in times of ever-tougher environmental regulations. A continuous deionization plant may also represent an attractive option for a situation where there are variations in feed-water TDS which can present difficulties in design and operation of ion exchange columns. Continuous deionization units are increasingly being applied in pure-water systems in pharmaceutical and microelectronics production facilities. In favourable circumstances (such as when utilised in combination with an upstream RO unit), CDI

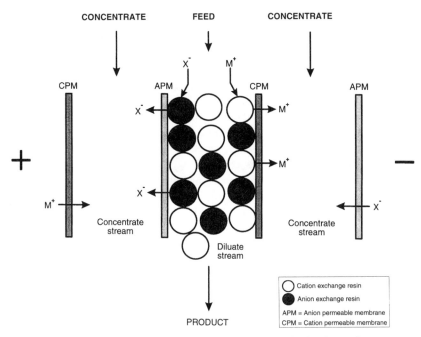

FIGURE 10.15. Schematic representation of a continuous deionization unit.

can produce water around the 10 megohm-cm purity level. In the pharmaceutical sector, CDI has been replacing conventional ion exchange in many recent installations. Its capacity to do the same in high purity systems in microelectronics plant (as opposed to its application in tandem with ion exchange) depends on the probably unlikely scenario of the CDI process being able to routinely deliver 18 megohm-cm water.

REMOVAL OF ORGANICS

There are a number of ways to remove dissolved organics from water. Appropriate methods are described below and it is worth stating that, in most pure-water systems, effective strategies will not usually involve reliance upon one single unit process for the elimination of organics.

Ion Exchange

This is feasible because many of the organic species present in water (especially in surface water) exist as large complex ions which can be removed by passage through ion exchangers. However, the choice of resin is important. The normal anion-exchange resins used for demineralization (described earlier) can suffer damage due to essentially irreversible adsorption of organics in the pore structure of the resin. Thus, for effective 'organic scavenging', specially tailored, macroporous ('macroreticular') resins have been developed from which the organic molecules can be removed during the regeneration step. In practice, reductions of 60%–70% of organics are usually obtained. Regeneration usually employs a hot (say 40°C) solution of sodium chloride and sodium hydroxide.

Reverse Osmosis and Nanofiltration

RO is said to be capable of achieving rejections in excess of 90% for organic substances with molecular masses above about 100, although the actual rejections achieved in practice does depend upon the type of membrane material and module employed. Indeed, this additional feature of RO (as well as desalting) is often made use of in high-purity water systems. However, RO should only be so utilized for the removal of residual organics since delivery of a feed containing a high organic loading to a RO unit may result in severe fouling of the membrane and a consequent serious decline in performance.

Nanofiltration is also effective for the removal of organic substances with molecular masses greater than about 200–400 and, in recent times, has been increasingly used by water-supply companies to treat water containing high levels of humic substances derived from the decay of vegetable matter which impart 'colour' to a natural water.

Activated Carbon Filters

The removal of large amounts of organic impurities by ion exchange or RO does not represent optimum practice—the fouling susceptibility of RO membranes has already been mentioned. Consequently, other means are required for the upstream removal of the bulk of the organics from a feedwater containing a high loading of such substances. This duty often employs the use of activated carbon beds. These consist (see Figure 11.3 of Chapter 11 of this book) of a vessel containing carbon granules which have been specially prepared from, for instance, anthracite, in a process which creates a vast network of micro and macro pores in the carbon granules. The macro pores are large capillaries, with diameters greater than about 5×10^{-8} m, that extend all through a particle and the micro pores, 10^{-9} m diameter upwards, generally branch off the macro pores and contribute a major part of the internal surface area of the activated carbon. The surface area can reach 1000 m^2/g and is crucial for conferring a high capacity for adsorption of the organics as the water percolates through the carbon bed.

The detailed mechanisms of selective adsorption of organic impurities are tied up with certain matching characteristics between the activated carbon and the organics; namely, a non-polar, hydrophobic nature. Charged ions and polar molecules tend not to be absorbed to any great extent on the activated carbon because of a high degree of attraction they have for the polar water molecules which tend to hold them in solution and oppose their adsorption on the carbon. The water molecules themselves tend also to be repulsed from the carbon surface.

Activated carbon can remove a broad spectrum of organic compounds but is much more effective in stripping those of low molecular weight because high molecular weight compounds tend to be poorly adsorbed. Chlorine is also effectively removed by activated carbon. In use, the carbon gradually loses its adsorptive capacity and eventually has to be either replaced by new material or rejuvenated to remove the adsorbed substances.

Other Methods

There are a number of related methods of removing organics by their decomposition by oxidation with the ultimate aim of the final product of the organics' decomposition being carbon dioxide. These techniques include subjecting the water to UV radiation (usually at 185 nm) or other similar oxidation methods employing ozone either alone or in combination with UV light or hydrogen peroxide.

Finally, for completeness, it should be pointed out that electrodialysis also exhibits a capacity to remove a proportion of the organics from water on account of the charged nature of some of the organic substances.

REMOVAL OF PARTICULATE MATTER

In addition to one of the basic objectives of a high-purity, water-treatment system to deliver a final product water free from suspended matter, there is a need to remove most of the solid burden in the early stages of water treatment in order to protect downstream equipment (such as RO units) from fouling or physical damage. Even colloidal substances, having diameters in the range 0.005–0.2 μm and lying at the interface between particles and dissolved species, need to be removed to protect RO membranes against rapid fouling by them.

Filtration Using Sand and Other Granular Media

These are designed to remove relatively coarse particles—perhaps down to the size range 10–30 μm. The use of sand as a filter medium has long been established but the process is much improved in efficiency if it employs filter materials with a range of sizes in successive bands in the filter vessel, the most coarse material being at the top (see Figure 11.2 of Chapter 11). These multi-media filters often contain silica sand, garnet sand and anthracite.

In these, all the water passes through the filter ('dead-end' mode) and the filtration mechanism largely involves 'depth filtration'—the particles being removed on a size-graded basis as they pass through a filter bed in which the pore size becomes progressively smaller.

In order to boost the effectiveness of the filtration, the water can be dosed with a chemical coagulant prior to the passage to sand- or multi-media filters. Alum (aluminium sulphate) is a long-established coagulant which at appropriate pH levels forms a fine precipitate of aluminium hydroxide into which finely divided matter (e.g. colloids), becomes incorporated. Polyelectrolytes, i.e. polymers that contain ionizable groups along their chains, are also widely used as coagulant aids. The coagulation process is assisted by the negative charge present on many colloidal species—enabling them to be attracted to opposite charges on the coagulant substances.

Cartridge and Microfiltration

These are suitable for removal of the finest particulates (typically those in the range 40 μm down to perhaps 0.1 μm) after the bulk of the suspended solids have been removed by the methods described above. Cartridge filters utilize disposable or cleanable filter media and are constructed in cylindrical or tubular elements with diameters typically around 70 mm outside and 35 mm inside, and lengths anything between 100 and 1300 mm. A wide variety of materials, including cotton, rayon, fibreglass and a range of other synthetic polymers, are used for disposable cartridge filters and the porous media in re-usable cartridges can be stainless steel, monel, ceramics or fluorinated-hydrocarbon polymers.

Microfiltration ('MF') units are essentially surface filters in the form of membranes with pore sizes in the range 0.1–10 μm. A wide variety of polymers are used for the manufacture of MF membranes: cellulose acetate, cellulose nitrate, polytetrafluoro-ethylene ('PTFE'), polyvinylidene fluoride ('PVDF'), polyvinyl chloride ('PVC'), polypropylene, polyesters, polycarbonate, nylons. Also inorganic MF membranes are available—including ceramics such as alumina which are especially useful for high-temperature applications. MF processes often operate in the 'dead-end' mode but are also available as cross-flow units in which a small proportion of the feed does not pass through the filter but is made to flow across the filter surface sweeping away the retained particles in a 'reject steam'. Most MF units are tubular with tube-diameters typically of around 5 mm, tube lengths about 1 m giving total membrane surface areas in the range 0.05–5 m². However, other MF module configurations are available—including some utilising sheet membranes and hollow fibres. An example of the latter type is equipment utilising hollow fibres with internal diameters of 0.3 mm, assembled in modules containing up to 20,000 m of fibre length, feed supplied to the outside of the fibres, operated in the dead-end mode with intermittent air 'backwash'.

Ultrafiltration

This technique utilizes membranes capable of removing undissolved solids, including colloids, together with bacteria and viruses down to a size of about 0.005 μm. However, water molecules, with an effective diameter of about 0.0002 μm, can pass through ultrafiltration (UF) membranes as can ions, low-molecular weight organics and gases.

UF membranes have pore diameters in range 10^{-3}–10 μm and can be made from a variety of polymers (including polyamides, polycarbonate, polysulphone, PVC, acrylonitrile, cellulose acetate) as well as some ceramics. UF membranes and modules bear many similarities to RO equipment. For instance, most UF membranes are 'asymmetric' but with active layers which are much thicker (about 10 μm) and more porous than RO membranes. UF membrane modules are available in tubular, flat sheet, spiral wound and hollow-fibre forms and utilize the cross-flow option. The rate at which the water molecules permeate an UF membrane depends not only on the pore size and tortuosity, but also on the transmembrane pressure—which is usually in the range 2–5 bar. The operational pressure in UF is considerably less than for RO. This is due to the fact that all the Δp in UF is available for pushing water through the membrane (i.e. no significant osmotic pressure effect) and also because of the less tight membranes utilized for UF which facilitate water permeabilities two orders of magnitude greater than attainable with RO.

Both UF and MF units are occasionally used as prefilters to RO units and are also applied at the terminal stage of a high-purity water line to remove fines and bacteria.

REMOVAL OF BACTERIA

Some aspects of the bacterial loadings in supply waters and the importance of their elimination are discussed in Chapter 11 of this book. Here, the available sterilization methods are considered. First, though, it is instructive to emphasize that a number of the unit processes used in the production of high-purity water themselves represent effective breeding media for bacteria. Thus, filters, resin beds, and activated-carbon columns provide large surface areas of high porosity that are ideal for microorganism attachment and the passing water supplies food for what can be exponential growth rates.

Chemical Dosing

Numerous chemicals are used as biocides in industry. Those relevant to pure-water systems are discussed below.

Chlorination Since chlorination is still the most common way of disinfecting public water supplies, it is likely that most feedwaters to a high-purity water train will contain a small residual chlorine content. It is possible that the feed on entering the plant may be subject to additional chlorination (or dosing with closely related oxidizing substances such as bromine and iodine). Chlorine can be dosed to a water either in gaseous form from cylinders or as an aqueous solution of sodium hypochlorite; these are chemically identical in terms of the final form of the 'chlorine' in the dosed water.

In fact, water is frequently dechlorinated as it passes through a pure-water line. This may be incidental if activated carbon filters are located near the front end of a system for removal of organics because activated carbon also eliminates chlorine. Alternatively, dechlorination may be intentional in order to protect downstream equipment (e.g. RO units with polyamide membranes) from damage by chlorine—in which case final dechlorination may be accomplished by dosing with a suitable chemical such as sodium bisulphite.

Ozone This is another effective biocide (and a powerful oxidizing agent). An advantage for use in a high-purity water system is that its use avoids chemical injection into the system since ozone (O_3) decays to oxygen. However, since this decay process proceeds quickly (minutes rather than hours), ozone cannot be used to confer long-term disinfecting powers. Ozone kills bacteria within seconds (considerably faster than does chlorine) and is said to be effective against viruses and pathogens. Ozone is generated by corona discharge, i.e. by the application of a high-voltage electrical discharge through a dielectric containing flowing air or oxygen. Dose rates of ozone for disinfection may be in the range 0.1–0.3 ppm. As mentioned earlier, ozone will also decompose organics but, for such purposes, higher dose rates may be necessary (depending upon organic loading). For organics, elimination ozone can be combined with UV irradiation of moderate intensity. If considered necessary, after the ozonation treatment the residual ozone can be removed by the use of high-intensity UV radiation which converts the ozone to oxygen.

Hydrogen peroxide The main use of this compound, H_2O_2, in many industries is for periodic sanitization of equipment.

Ultraviolet (UV) Irradiation

This represents the most common method for the final sterilization of water near the terminus of a pure-water system. UV radiation is that part of the electromagnetic spectrum that is just on the short-wavelength side of the visible spectrum and ranges from 100 to 400 nm in wavelength. The attributes of the use of UV irradiation for the final sterilization of water are that no chemicals are injected into the water—also avoiding the storage and handling of chemicals, simple, compact, low-maintenance, low-energy equipment utilized and low contact time—avoiding the space and cost of providing a contact vessel. Against these features the method does not confer any residual disinfecting power to the water which is hence vulnerable to bacterial contamination downstream of the UV sterilizer. Additionally, the presence of any suspended matter may block the UV radiation from pockets of water passing through the sterilizer; however, this is not usually a problem near the end or a high-purity water line. Another consideration is that bacteria are usually converted to pyrogens which may require subsequent fine filtration (possibly ultrafiltration).

The sterilizing efficiency of UV radiation varies with wavelength and is at a maximum at around 260 nm, which is lethal to bacteria, viruses and a wide range of other microorganisms. Hence low-pressure mercury lamps, which have power consumptions typically around 65 W and which produce a strong, almost monochromatic, emission at 254 nm, are widely utilized for disinfection purposes, although medium-pressure lamps, with a broader wavelength spectrum and higher-power outputs, are also effective. The UV equipment (see also Chapter 11) usually consists of lamps encased in quartz tubes which have a high transmissivity for UV radiation

Heating

This well-known method of disinfection has represented the traditional approach for water used in the pharmaceutical industry and it has been an important factor in the retention in that industry of distillation for the removal of ionic impurities since distillation is seen as an automatically sanitizing process. In some micro-electronics and pharmaceutical factories, high-purity water lines are run at elevated temperatures in order to control bacterial levels. For instance, water for injection is often continuously circulated between a storage tank and points of use at temperatures of 80°C and above and, if distillation has not been used for demineralization, the required heating will involve heat exchangers. Finally, steam is the accepted method of periodic sanitizing of equipment in the pharmaceutical industry.

Membranes

Reference to Figure 9.1 indicates the potential of membrane filtration for disinfection purposes. Indeed, 0.2 μm membrane filters are often used for capturing bacteria. Reverse osmosis, nanofiltration and ultrafiltration membranes possess a high degree of disinfection action. Ultrafiltration will be least effective in this respect. Nanofiltration is reported to be effective for the removal of bacteria and viruses but cannot be relied on for the elimination of pyrogens. Reverse osmosis, having the tightest structure, represents the most efficient disinfection device of these three membrane processes, but even its absolute reliability for this function is not universally accepted. Moreover, for reasons recounted in relation to other particulates earlier, the supply of water containing a high biological burden to a RO module will result in rapid decay in performance. Also, in microelectronic factories water issuing from a RO unit usually undergoes further treatment (such as passage through ion exchangers and fine filters) where further bacterial growth is possible; hence final disinfection by, say, UV radiation is invariably necessary. In the pharmaceutical sector, the regulation regarding disinfection processes vary in different countries. The Japanese Pharmacopoeia approves both ultrafiltration and reverse osmosis for the production of water for injection (WFI). The U.S. Pharmacopoeia authorises reverse osmosis but not ultrafiltration for such purposes. In Europe at present, membrane processes are not acceptable for the production of WFI disinfection but this matter is subject to a current debate that may lead to a change in the European Pharmacopoeia regulations in the near future.

CONCLUDING REMARKS

The objective of this chapter has been to provide an appreciation of the wide variety of techniques available for the removal of the different types of impurity in feedwaters. This is important because of the variable practices adopted by different users of high-purity water. These differences arise not only on account of varying requirements for the final product but, more importantly, because of the varying constitution of feedwaters. A further substantial complication is that virtually never can one type of impurity be removed efficiently by one unit operation. Thus, dissolved ions require a 'roughing demineralization' treatment (for which distillation, RO and ED are candidate processes) and a subsequent 'polishing stage' (which involves ion exchange). Similarly, as has been briefly discussed, dissolved organics, suspended solids and microorganisms are invariably removed in several stages by different processes. Thus, a study of any one high-purity water system, however soundly chosen, does not reveal the entire picture as there are usually a number of options open to a specifier or designer of a high-purity water system and a knowledge of the wide range of potential techniques available is an essential prerequisite for an informed decision on system design.

11 The Design of an Ultra-Pure Water System for Use in the Manufacture of Integrated Circuits

R. GALBRAITH

INTRODUCTION

Ultra-pure water is used in the semiconductor industry principally to remove contaminants from the surface of wafers being processed. It is used in a variety of cleaning stages following processes such as etching. This is a process step during which the wafer is immersed for a specified time in concentrated acid, e.g. sulphuric/hydrogen peroxide, hydrofluoric, etc. Residual acid on the wafer surface must be removed quickly to prevent over-etching and whatever chemical is used to achieve this must not itself contribute any contaminant to the wafer surface.

The chemical normally used for this process is ultra-pure water (UPW), chosen because it is readily available in crude form, is relatively cheap, can be purified to a state where it contains virtually no contaminants, and is an excellent solvent for most of the chemicals used in wafer processing. However, the excellent solvent properties of UPW also make it difficult to maintain in its pure state due to its affinity for most contaminants existing in the atmosphere, in tanks, vessels and pipework.

IMPACT OF CONTAMINATED WATER

It is worth putting into perspective the potential impact that a contaminated water supply could have in the manufacturing area. Unlike other process chemicals, ultra-pure water is used throughout the wafer fab and accounts for greater than 99.5% by volume of all chemicals used.

The contaminants that are of concern can be loosely classified into five main categories:

- Ionic
- Non-ionic
- Organic

Cleanroom Design. Edited by W. Whyte © 1999 John Wiley & Sons Ltd

- Bacteria
- Dissolved gases

The looseness of the classification is due to the fact that some of the contaminants may be considered to be in more than one class, depending on conditions.

Ionic

Ionic species form the largest percentage of all contaminants which must be removed from the source water. The concentration and types that are present depend on the source of the local supply but the common ions such as calcium, sodium, potassium, magnesium, chloride, sulphate, carbonate and bicarbonate will almost certainly be predominant. The principle methods for removing ionic species are reverse osmosis and ion exchange, economic considerations determining the extent to which each is utilized. Other potential sources of ionic contamination are regeneration chemicals, on-line chemical dosing, pumps and pipe systems.

These contaminants, if present, will adversely affect the semiconductor's performance and will almost certainly result in its failure. Ionic contaminants have traditionally been the main area of interest in terms of removal, partly because the methods are well established and partly because the means of monitoring the removal efficiency have been readily and economically available. The unit used for measuring ionic contamination is resistivity, which is the inverse of conductivity but an easier unit to use at such low levels. Resistivity is quoted as megohm/cm at a reference temperature of 25°C, this being important as resistivity is temperature dependent. Figure 11.1 shows the relationship between resistivity and ionic contamination as ppm sodium chloride.

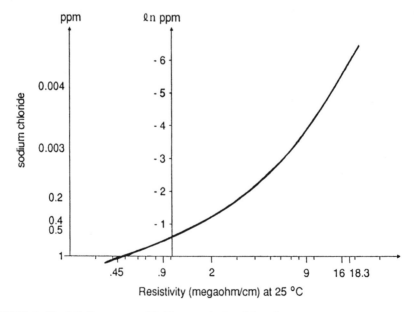

FIGURE 11.1. Resistivity vs. ppm (NaCl or equivalent) for ultra-pure water.

The theoretical limit of resistivity is not infinite, as the water molecule is weakly polar, and is in fact 18.27 megohm/cm at 25°C. Modern UPW plants produce water exceeding 18 megohm, i.e. the equivalent sodium chloride concentration would be less than 0.004 ppm.

There are also a range of ions which are only weakly ionic and their presence will also have a bearing on the technology chosen for a particular plant.

Non-Ionic

This class contains contaminants which are either truly non-ionic, being formed of covalently bonded materials, or so weakly ionic that they may prove difficult to remove using standard ion exchange resins. They may also be in the form of insoluble particulate matter which is present in the source water or which has been generated by the water treatment plant, e.g. resin fines.

These contaminants may cause yield loss due to distortion between successive layers on the silicon wafer or by short circuiting. They are generally specified in terms of particles per litre above a particular size. The size of particle which can be tolerated is a function of the device geometry but would normally be less than 0.1 μm.

Organic

The amount of organic compounds present in source water depends on whether it is from a hard water catchment area or from a surface water reservoir. Hard water would characteristically be clear, with a Total Dissolved Solids (TDS) concentration of over 500 ppm. Surface water reservoirs, on the other hand, tend to produce soft water which has a TDS of below 100 ppm but which can be heavily laden with organics leached from the soil of the surrounding catchment area. The organic compounds in soft water are complex and consist of humic, fulvic and other organic acids created principally from decaying vegetation. Organic contamination can be difficult to remove and the type and concentration can vary considerably subject to seasonal changes. Organic molecules can cause yield loss due to staining and corrosion on the wafer surface, or by conversion to smaller organic species by some of the water-treatment processes. They can also cause fouling on ion exchange resins and membrane filters thus reducing the ability of these stages to remove other contaminants. Where this fouling is irreversible, it can result in premature failure and consequential replacement of resins and membranes.

Organic concentrations in UPW should be less than 10 ppb and preferably below 5 ppb. On-line instrumentation at various stages of the plant is essential to provide early warning of problems and to ensure that specifications are achieved. Total organic concentration can be measured by conversion to carbon dioxide using either UV radiation or chemicals, or by a combination of both.

Bacteria

Bacteria are ubiquitous, capable of existing in either aerobic (with oxygen) or anaerobic (without oxygen) conditions and surviving in extremely harsh conditions to form resistant strains. In UPW systems they are usually of the Gram-negative type and in terms of effective particle size can be sub-micron. The components used in water purification, such as ion exchange resins, provide an ideal environment for bacteria to thrive and once established they can be difficult to remove as the chemicals traditionally used would themselves cause contamination and irreversible damage to the substrate.

Bacteria can contaminate semiconductors by either behaving as a particle or by breaking up and leaving a number of ionic species and pyrogens on the wafer surface.

Dissolved Gases

The most common dissolved gases in water are oxygen and carbon dioxide and this varies depending on the source, pH and water temperature. Dissolved oxygen can cause enhanced oxidation at the wafer surface and, unless removed, will be a variable in the final supply water. Dissolved carbon dioxide forms bicarbonate and carbonate ions in solution, the equilibrium being dependent on pH.

Both gases are readily removed and the methods used will be discussed later in the chapter.

To achieve acceptable wafer yields, all the above contaminants must be removed to extremely low levels and consistently maintained. The level of each contaminant which is considered to be acceptable is being constantly revised (but never upwards) as device geometries become smaller and therefore less tolerant of impurities.

Table 11.1 gives a list of common impurities and the typical level of control required in a modern plant.

TABLE 11.1. Specifications for final water quality.

Parameter	Specification
Resitivity at 25° C	>18.0 megohm
Total organic carbon (TOC)	<5 ppb
Silica (dissolved)	3 ppb
Particles (>0.1 μm)	<100 particles/ litre
Bacteria	< 5/ litre
Cations (Na, K, Ca, Mg, NH_4)	< 0.1 ppb (each ion)
Anions (F, Cl, SO_4, NO_3, Br, PO_4)	< 0.1 ppb (each ion)

PLANT DESIGN

Source Waters

Source waters can be broadly classed as hard (high TDS) or soft (low TDS) and each provides different challenges to the UPW plant designer. The basic principles and removal mechanisms are common to both source types but their implementation and even position in the process stream is specific for each installation.

The two fundamental pieces of information which will initially influence the designer's choice of technologies are:

1. *Source water quality.* It is essential to gather as much information as possible on the source water supply. This may be obtained from the local water authority

or from other large users in the same area and should encompass not only the seasonal variations but also the frequency of abnormal conditions. There may already be a plant on the same site which can provide a detailed history of operational problems and successes over many years.

2. *Final product quality*. It is necessary to assess what water quality is required at the wafer surface to satisfy today's specifications and to determine if the plant is capable of adapting to and achieving future industry standards.

The next factor that will influence the designer is the capital and operating costs of the plant. With the construction of a new wafer fab now around US$1 billion, the capital available is tightly controlled and the water treatment plant will have to compete for funding against other utility costs and process equipment. The operating cost will directly contribute to the die manufacturing cost and must therefore be minimized if the manufacturer is to remain cost-competitive. The following is an overview of the most common technologies used in the semiconductor industry.

Pre-Treatment

The term pre-treatment implies that this is a conditioning stage for something that will follow later and in most plants it is the precursor to resin and membrane separation techniques. However, its importance should not be underestimated as its success or failure will strongly impact the performance of the entire plant. There is a large selection of pre-treatment processes and variations within each category. These include filtration, flocculation, dechlorination, organics removal, softening, degassing, disinfection and temperature adjustment.

Multi-Media Filter

A typical multi-media filter as shown in Figure 11.2 consists of a steel vessel with a rubber lining. The service flow is downwards through a packed bed ranging from a coarse anthracite layer at the top to a fine sand layer at the bottom. It is a coarse depth filter used to remove particulate matter from the incoming water but is inefficient at particle sizes much below 5 μm. Where appropriate it can also serve to remove precipitates which are the product of chemical injection upstream of the unit.

The unit is normally cleaned by reversing the service flow and fluidizing the bed to remove any accumulated matter. The frequency of cleaning can be dictated by a set batch volume or when the pressure drop across the unit reaches a pre-set value. Periodic chemical cleaning of the unit to remove fouling may also be required if the normal service backwash is incapable of removing all the captured material by agitation alone. A variation which may be incorporated into the normal cleaning cycle is an air scour to aid in dislodging particles from the bed, especially if flocculation has preceded this stage.

Activated Carbon Filter

As can be seen from Figure 11.3, this filter is of similar construction to the previous unit but contains activated carbon. It is used principally to remove chlorine which is

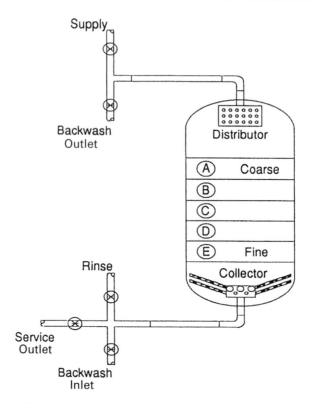

FIGURE 11.2. Multi-media filter.

damaging to the thin film membranes commonly used in reverse osmosis (RO) systems. Activated carbon is also capable of removing low molecular weight organic molecules which may penetrate some of the downstream units.

Normal service cleaning of this unit is similar to the multi-media filter but at a lower backwash flow. The organic molecules removed during service are not removed by backwashing but are instead retained within the pore structure of the carbon granules. The removal efficiency eventually decreases to a level at which the usual course of action is total replacement of the carbon. The exhausted carbon may, in some circumstances, be returned to the manufacturer for regeneration, but this is generally not economic.

Organic Trap

This unit, shown in Figure 11.4, consists of a rubber-lined vessel, but in addition to the usual distributor and collector there is an injection system to introduce regeneration chemicals (normally brine). The capital and operating costs of this unit are higher as it contains ion exchange resins selected for their ability to capture large organic molecules. The resin chosen must also be capable of eluting these organics during the regeneration cycle or it will otherwise become irreversibly fouled. In practice the efficiency of macroreticular organic scavenger resins is typically 60%–70% and with proper care this performance can be maintained for many years.

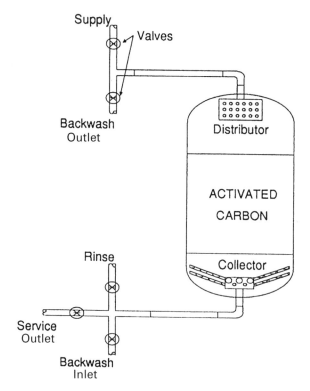

FIGURE 11.3. Carbon filter.

In some instances a softener resin may also be included within the same vessel to remove hard calcium and magnesium ions which may cause scaling problems when concentrated within an RO system.

Chemical Addition

The need for chemical injection varies widely and in some applications may not be required at all. Among the main objectives of chemical dosing are the following:

1. Precipitation and coagulation of species which will cause fouling in later stages.
2. Bacteria control by disinfection with chlorine or ozone.
3. Prevention of scaling within RO systems.
4. pH adjustment
5. Dechlorination

Chemical addition must be carefully controlled otherwise the chemicals which have been introduced may themselves present problems farther downstream.

Filtration

The final conditioning stage which is normally included upstream of an RO system is filtration using disposable cartridge filters. This is done to protect the sensitive mem-

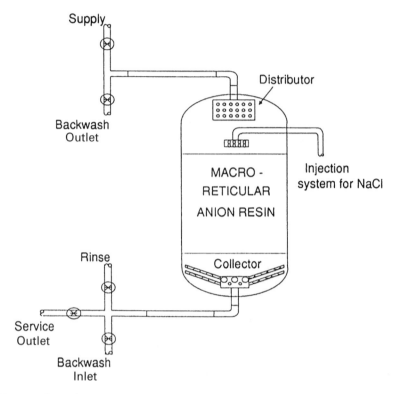

FIGURE 11.4. Organic trap.

brane surface from physical damage by particles originating from the media in the pre-treatment units. A filter with a pore size of about 1 μm can provide sufficient protection without adding substantially to the plant operating costs.

It is worth noting also that there may be some temperature adjustment incorporated in the pre-treatment stream. The temperature of many water supplies varies seasonally and this has an effect on the efficiency of some processes and significantly affects the flux rate of an RO system. Even if a seasonal variation in water can be tolerated within the plant there may be a requirement from the end-user for temperature stability to allow close control of process conditions, e.g. semiconductor manufacture. The cost of maintaining a constant temperature is dependent on the amount of heating or cooling required, but for the former it may be possible to use waste heat to achieve the desired temperature without incurring high fuel costs.

Reverse Osmosis

The detailed theory of reverse osmosis has been explored in Chapter 10 of this book. However, it is worth briefly mentioning a few of the basic principles and terminology:

Osmotic pressure. This is a function of the concentration of the dissolved solute (or in other words, as the water gets dirtier the osmotic pressure rises).

Flux rate. This is the rate at which the product water flows through the membrane and is a function of the difference between the pressure applied by the RO pump and the osmotic pressure. Therefore, as the water gets dirtier, either the product flow rate will be reduced if the output pressure from the pump is fixed, or the supply pressure must be increased (at increased energy cost) to maintain the same product flow rate.

Sensible osmotic pressure. The membrane is affected by the osmotic pressure adjacent to the membrane surface. Unless great care is taken, this may be far higher than the main stream osmotic pressure as the passage of pure water to the product side of the membrane leaves the concentrated impurities behind. This means that unless good mixing occurs between this boundary layer and the bulk of the fluid, a large concentration difference or polarization will occur.

The conclusion to be drawn from the above three points is that there should always be a high Reynolds number at (and hence liquid velocity across) the membrane surface. This will influence the configuration of the membranes in any given system but the system design should always satisfy the membrane manufacturer's design for each individual element.

Dirt passage. No RO membrane acts as an absolute filter and there is always some dirt passage. This partly explains why the RO unit is operated at a pressure far in excess of the osmotic pressure as this helps to optimize product quality. The other reason is that the higher the flux rate the smaller the plant size, and capital cost, for a required output.

Membrane materials. These are available in three broad categories i.e. cellulose acetate, polyamide and thin film composites. Selection of the optimum membrane type depends on the quality of the feed water at the RO system inlet and should take into consideration the effect on the membrane of any chemical dosing. Each type of membrane offers different advantages (and disadvantages) and some installations have attempted to optimize system performance by using different types of membranes on the first and successive stages.

Membrane configuration. There are basically two principal membrane configurations available for use in UPW plants. The first is hollow fibre and this type is generally made from polyamide. The fibres look like human hair and millions of them are fastened together in a unit with the result that each unit has a very large active surface to volume ratio. Reasonably compact plants can therefore be installed but the principal disadvantage is the interstitial spacing of the membranes. One of the contractual constraints placed on users by membrane manufacturers is a maximum silt density index (SDI). This is a measure of how quickly the feedwater will block a 0.45 μm filter and therefore the fouling potential of the membrane. If the water source has a high organic content, then the capital and operating cost of pre-treatment is prohibitively expensive and hollow fibre systems are rarely used. They are, however, generally suitable for use with sea water or very hard waters.

The second and more favoured configuration is referred to as 'spiral wound' in which a membrane material and supports are rolled into a 'jam sponge' assembly. Compared with hollow fibre, this arrangement has a far lower surface area to volume ratio and therefore the most common materials used are thin film composites and cellulose acetate. Both of these membrane configurations are described in Chapter 10.

The RO system concentrates the impurities present in the feed water into a reject stream and allows passage of purer water as product. The arrangement shown in Figure 11.5 is a 2:1 configuration common in spiral wound systems, i.e. the reject streams from tubes 1 and 2 are fed to a second stage for further treatment. Each of the tubes in the example shown contain six interconnected RO elements, but there are a number of alternative configurations which can be used.

Unlike the pre-treatment units (and those farther downstream) which only create waste during the regeneration cycle, the RO system produces a continuous waste stream. The operating cost of a system is a balance between two main objectives:

1. A reduction of waste stream volume by maximizing the percentage recovery across the system (which may be typically around 75%–85%). This reduces the cost of supply water, feedwater chemicals and effluent disposal costs.
2. To obtain the maximum life from the membrane and the on-line system availability. If the percentage recovery of the system is too high, or if the feedwater is not properly conditioned, the system will require more frequent maintenance cleans and possibly premature replacement of the individual elements.

The performance and physical characteristics of a typical RO element used in industrial applications are given in Table 11.2.

The product water from the RO system is transferred to a storage tank which is blanketed with nitrogen to inhibit the growth of aerobic bacteria. This may be a stand-alone storage tank or may be an integral part of another (optional) treatment stage such as degassification or dearation.

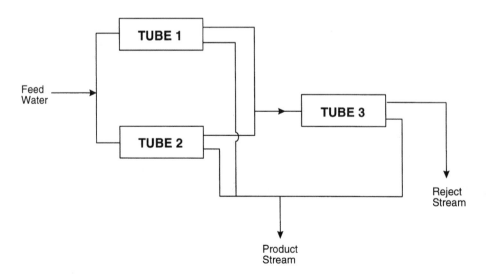

FIGURE 11.5. Reverse osmosis unit (2:1 configuration).

TABLE 11.2.

Length	1015 mm
Diameter	203 mm
Max. pressure drop per element	10 psi
Operating temperature	5–45° C
Feedwater: Silt Density Index	< 5
Applied pressure	about 15 bar
pH operating range	4–11
Water recovery per element	10%
Permeate flow	32 m³/day
Chloride ion rejection	99 %

Deaeration/Degassification

In some plants the stage immediately following RO is the removal of dissolved gases. This may either be on economic grounds to reduce the operating costs of downstream ion exchange beds by removing CO_2 or due to a process requirement specified by the end-user, e.g. low dissolved oxygen content.

The removal of CO_2 can be achieved cheaply by adjusting the pH and blowing air upwards through a packed column as the water cascades downwards. Great care must be taken to filter the air to avoid contamination by particles or by organic species.

The removal of dissolved oxygen can be achieved in two ways. The most commonly chosen option is vacuum degassification, which removes oxygen and other dissolved gases by reducing the partial pressure. An alternative is to use a tower similar in construction to that for CO_2 but using an oxygen-free gas to strip oxygen from the process stream. This option may be feasible if a site has an abundant supply of cheap nitrogen, as it has the benefit of being a virtually maintenance-free system.

In each of the above cases the collection sump at the bottom of the column can act as the RO storage tank. However, these are usually of limited capacity and may act only as an intermediate stage if there is a requirement to store a large volume of RO quality water.

The water quality produced by the RO system is, by general standards, a clear pure water, but by the semiconductor industry standards it is unacceptable for wafer processing. The required quality can only be achieved by further purification methods, principally demineralization through ion exchange.

Demineralization

The process of demineralization relies on the principle that ion exchange resins have different affinities for the ions most commonly found in source waters. Positively charged ions such as sodium are referred to as cations and negatively charged ions like chloride are called anions. A water softener, which may be installed in the pre-treatment stage, functions by exchanging 'soft' sodium ions held on the resin for 'hard' calcium and magnesium ions in the process stream. The difference with the ion exchange resins used at this stage of the plant is that the contaminant ions in the inlet water are exchanged, not for other contaminant ions, but for pure water. Although this is now usually achieved in a single unit containing cation and anion resins, the mechanism by which this is achieved is more easily explained by considering the two types of resin separately.

Cations   A cation unit in its simplest form is shown in Figure 11.6. The unit consists of a rubber-lined steel vessel which has service and regenerant inlets and outlets and strainers to contain the small resin beads. There is a wide selection of cation resins available and the type and quantity of the resin selected is dependent on the type and number of cations remaining in the process stream. There are some fundamental rules that govern the behaviour of ion exchange, which we will now consider.

1. Resins have a higher affinity for multivalent ions than they do for monovalent ions, e.g. a cation resin bead would prefer to have a calcium ion which has a double charge to a sodium ion with a single charge.
2. Where the ionic charge on two ions are equal, the resin has a higher affinity for the ion with the higher molecular weight, e.g. a cation resin would prefer to have a sodium ion to a hydrogen ion.
3. The exchange process is reversible and the equilibrium position established based on the first two rules can be shifted by a concentration effect.

The direction of flow during the service mode is downwards through the unit. A cation resin will release the hydrogen ions attached to its surface in exchange for other ions:

$$R\text{—}H^+ + Na^+ = R\text{—}Na^+ + H^+$$

$$2\,(R\text{—}H^+) + Mg^{2+} = 2R\text{—}Mg^{2+} + 2H^+$$

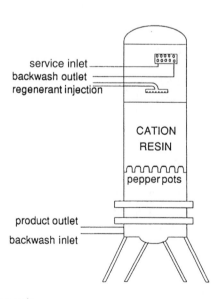

FIGURE 11.6. Cation unit

The resin will also release a sodium ion captured previously in preference for a magnesium ion:

$$2(R\text{---}Na^+) + Mg^{2+} = 2R\text{---}Mg^{2+} + 2Na^+$$

The service cycle therefore consists of a number of equilibrium reactions that occur down through the cation bed. The net effect is that a layer rich in multivalent ions is established at the top of the bed which progressively deepens with time, displacing lighter ions in the process. The unit eventually becomes exhausted when sodium ions are unable to find a hydrogen ion exchange site and break through. When this stage is reached the unit is removed from service and regenerated, generally with hydrochloric acid. The contaminant ions are driven back off the resin due to the concentration effect which occurs when a solution containing sufficient numbers of hydrogen ions is introduced to force the reverse reactions.

$$R\text{---}Na^+ + \text{excess } H^+ \Longrightarrow R\text{---}H^+ + Na^+ + \text{excess } H^+$$

Anions   The anion unit is similar in construction (Figure 11.7) and located downstream of the cation unit. The anion resin obeys the same general rules as for cations except that contaminant ions are exchanged for hydroxyl ions which are attached to the resin:

$$R\text{---}OH^- + Cl^- \Longrightarrow R\text{---}Cl^- + OH^-$$

$$2\,(R\text{---}OH^-) + SO_4^{2-} \Longrightarrow 2R\text{---}SO_4^{2-} + 2\,OH^-$$

The hydroxyl ions that are liberated combine with the hydrogen ions from the cation unit to form water molecules. Since the hydroxyl ions are immediately paired off, they

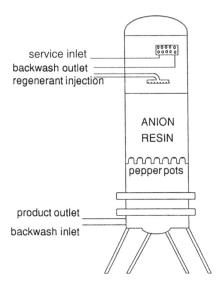

FIGURE 11.7. Anion unit.

are unavailable to participate in the reverse reaction previously discussed for cations and this results in a lower slippage of ions from the unit. When the anion bed is exhausted, it is regenerated with excess sodium hydroxide:

$$R—Cl^- + \text{excess } OH^- = R—OH^- + Cl^- + \text{excess } OH^-$$

The water quality from a cation–anion pair may, at best, achieve 10 megohm which is still not sufficiently pure for the semiconductor industry. The resistivity can be further increased by placing a second cation–anion pair in series but to achieve the desired product quality would require an impracticable number of such pairs. This problem is overcome in practice by intimately mixing both cation and anion resins in a single vessel and thereby effectively providing millions of cation–anion pairs.

Mixed Bed Units

A typical mixed bed unit, as shown in Figure 11.8, is fitted with a central distributor which is necessary if the resins are to be regenerated *in situ*.

Although the concept of a mixed bed appears to be a neat and economic alternative to separate cation–anion units there are some mechanical and chemical difficulties which must be overcome. The resins must be intimately mixed while in service but must be able to be separated when the bed is exhausted to allow regeneration. This is accomplished by selecting resins of different densities which separate when the bed is expanded during the initial backwash stage. When the upflow of water is stopped, the heavier cation resin falls to the bottom while the lighter anion resin remains on top. The quantities of resin used are calculated such that the interface between cation and anion coincides with the central distributor. The cation resin is regenerated by introducing acid into the bottom of the unit and out through the central distributor. The anion resin is regenerated in a similar fashion but by introducing caustic solution at the

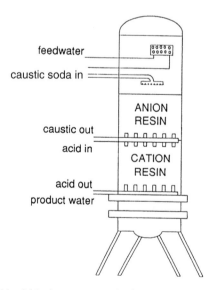

FIGURE 11.8 Mixed bed (during regeneration)

top of the unit and exiting through the central distributor. There are also rinse operations on each resin to remove excess acid and caustic from the respective parts of the bed. Re-mixing of the two resins is achieved by draining down the water level until it is just above the anion surface (the actual level is critical to its success) and then agitating with air or nitrogen. Draining down is essential to ensure that the resins do not re-classify once the air is shut off due to the difference in their terminal velocities. The unit is finally rinsed to remove any pockets of regenerants before being returned to service or placed on standby.

The main mechanical problem inherent in the mixed bed design is caused by the hydraulic forces placed on the central distributor not only during service but particularly when the bed, having been compacted during service, is lifted during backwash. The distributor must also be resistant to corrosion which could be experienced during the injection of regenerants.

A prerequisite of a good regeneration is a good classification of the resins following backwash. In practice some of each resin may end up on the wrong side of the distributor, will therefore be exposed to the wrong regenerant, and will consequently be totally exhausted when returned to service. Some causes of poor classification are:

- Breakage of cation resin beads, producing fines light enough to classify with the anion resin.
- Fouling of the anion resin, effectively increasing its density and allowing it to classify with the cation resin.
- Loss of resin due to attrition, which through time will shift the interface position away from the central distributor.

Even when a perfect classification is achieved there is inevitably some cross-contamination of the regenerants at the interface. This may result in some leakage of chloride ions (and to a lesser extent sodium) from the bottom of the bed when returned to service and will have an effect on product quality.

Polishing Mixed Bed

Although the conventional two-resin mixed bed produces a high purity water, the impurity levels are still higher than what is acceptable for applications such as semi-conductor processing. There are several options available to the plant designer to remove the few remaining contaminant ions from the process stream.

One option which may be considered is to install a polishing mixed bed in addition to the primary mixed bed described above. The structure of both vessels is similar, but the polishing mixed bed, as shown in Figure 11.9, includes a third resin which is inert and serves no function while the unit is in service. The function of the inert resin is to provide a barrier (normally 150–200 mm deep) between the cation and anion resins during regeneration and effectively prevent cross contamination at the central distributor. Its density must be carefully selected to ensure that it classifies between the cation and anion resin following backwash. The inert resin is expensive and has the disadvantage that it occupies bed capacity during service but it does introduce the capability of producing ultra-pure quality water.

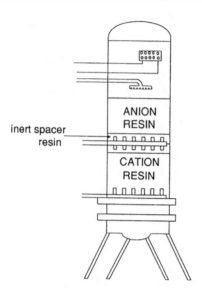

FIGURE 11.9. Polishing mixed bed (during regeneration).

An alternative approach is to install non-regenerable mixed beds downstream of the primary mixed beds. The advantages of this option are:

- lower capital cost due to simple vessel construction, smaller size and no requirement for chemical lines and valves;
- no potential source of leakage of regenerant chemicals into the final product water:
- lower initial cost of resin.

The disadvantage is that the resin must be discarded and replaced once the water quality begins to deteriorate as the resins are not specially graded and therefore cannot be separated for regeneration. However, if the inlet water is already very pure and long service times obtained, it is a viable proposition as the ongoing cost of replacement resin is favourable compared with the capital cost of a polishing mixed bed.

Ultraviolet (UV) Systems

The removal of bacteria from ultra-pure water is necessary because of the potential damage they cause if they reach the wafer surface. A single bacterium can act in a similar way to a particle, causing non-uniformity which can affect subsequent layers of wafers, and the ionic components within the bacterium can cause damage if the cell wall is ruptured. Some method of disinfection is required to control the growth of viable bacteria within a polishing loop and distribution system. However, a non-viable bacterium contains the same amount of organic and ionic material as a viable bacterium and must also be removed from the stream.

The most commonly used method of bacteria control is UV radiation followed by sub-micron filtration. This method is, however, only effective against the planktonic bacteria which are carried in the water stream and has no effect on the sessile bacteria population which are lodged in resins, filters, pipework and valves. The latter can be released in large numbers into the main stream if subjected to mechanical shock or sudden changes in flow rate.

An UV sterilizer kills bacteria by physical means in contrast to other techniques such as chlorination and ozonation which act by chemical attack. It functions by exposing suspended bacteria (planktonic) to UV light centred around a wavelength of 254 nm. Bacteria exposed to this wavelength are killed due to damage to their DNA and as they remain intact they can be easily removed by a particle filter. The efficiency of the unit depends on the type of bacteria present, the dosage applied and the contact time.

UV systems can be constructed either as single lamps or multi-lamp units. UV light is produced by a low-pressure mercury discharge lamp which is protected from the water by a quartz sleeve. Quartz is the normal material chosen as it has a high transmission of UV light and the loss of intensity is typically less than 5%. The unit is constructed to create turbulent flow which ensures that bacteria existing in colonies or attached to other particles are exposed to the radiation. The performance of the lamp is measured by a UV sensor as the intensity decreases during the lamp life, resulting eventually in lamp replacement after a period of about 6 months. It is normal practice to size the power of the lamp such that the power available at 50% transmission is still sufficient for bacteria control. Units should also be fitted with an hours run meter for maintenance operations and over-temperature protection.

The advantage of UV systems is that they do not introduce anything to the fluid stream which may have to be removed at a later stage. The disadvantage is that there is no residual effect and it is therefore incapable of killing organisms which are not circulated through the unit.

Another method increasingly used is ozonation at selected process stages, but it is not commonly used in the distribution loop due to the potential attack by ozone on items such as gaskets and seals.

Organic Reduction Using UV Radiation

Most UPW plants now use UV light centred at 185 nm to reduce organic contamination. The shorter, higher energy wavelength can cause oxidation of organics into water and carbon dioxide. The dominant mechanism is by formation of hydroxyl radicals which attack organic species. The other more direct mechanism at work is the breakage of certain organic bonds, but this is a slower reaction and is not effective on all carbon bonds. These units are installed upstream of ion exchange beds which are then able to remove the oxidation products more easily than the original compound.

Final Filtration

The last line of defence in the UPW system is filtration of particles which are large enough to damage a semiconductor. The general rule of thumb is that particles greater than one-tenth of the line width have the potential to cause damage. The most common type of particle filter consists of a stainless steel housing containing cartridge filters rated to 0.1 μm (or better) and 20 in. to 40 in. in length. Cellulose acetate filters have now been superseded by materials such as nylon, PVDF and PTFE and there is also the option of positively charged filters if desired. The filter elements are expensive, but as the particle challenge at this stage should be low, it is possible to get several years of life from them. Great care must be taken when replacing filter elements to ensure that particle contamination is not introduced into the distribution pipework. Chemical resistance of the filter material may also need to be considered if the plant operates a continuous or periodic sterilization programme. In general, the smallest absolute rating which can be tolerated in terms of pressure drop should be chosen and the design should take account of the need to change to smaller pore sizes to satisfy technology demands for smaller die geometries.

Ultrafiltration has gained popularity over recent years as a final filter and it also has the capability of removing large organic molecules. However, it is not universally accepted as a final stage by plant designers and is more frequently used in addition (and at high capital cost) to the standard cartridge filters. It is similar to reverse osmosis in that the impurities are continuously removed in a reject stream which is typically about 5–10% of the feed.

Polishing Loop

There are a number of factors to be considered when designing the polishing loop, and the major components within the plant have already been discussed. The selection of the distribution pipework is crucial because all the good work that has gone before can be ruined by a poorly designed distribution system. It is also relatively easy to change the type of RO membrane, resin or filters from the original selection, but this luxury cannot be afforded to the supply pipework without major disruption to the production areas. The most common materials used are fluorinated polymers and in particular polyvinylidene fluoride (PVDF). Although the traditional plastics such as PVC and ABS are commonly used in the front end of the plant, PVDF is preferred for the polishing loop for the advantages it affords, albeit at a higher cost. Its advantages are that it

- is fusion welded and requires no glues;
- minimizes bacteria growth;
- is resistant to sterilizing agents such as hydrogen peroxide;
- is an inert material which does not leach organic or ionic contaminants to the system (following initial rinse up).

While PVDF itself may not provide any nutrients that promote bacteria growth it is also important to ensure that water is continuously flowing throughout the system since bacteria are more likely to thrive in static conditions. The system should be designed to eliminate dead legs in the branches to process tools and to achieve a minimum velocity of 1 m/s in all sections. This may be theoretically possible but in practice it is not so easy, especially for branch lines. Wet processing equipment does not have a constant demand for water as the process is cyclic in nature and it is necessary to install a return line in each case to ensure continuous flow. This unused water is returned to the start of the loop in the central plant which is usually a high purity tank and then blended and recirculated with make-up water from the plant. Once the system has been designed, installed and is in operation it is not uncommon for wafer fabs to alter the equipment layout to satisfy business demands or to introduce new equipment having completely different service requirements, and in so doing alter the entire supply pattern. In such cases it may not be possible to re-design from scratch and some compromises have to be accepted.

Figure 11.10 shows a suggested layout of an ultra-pure plant. Some of the stages included can be omitted and there are other units that need to be added to deal with specific contaminants. However, if you were to walk into a UPW plant today, you would probably find something very similar.

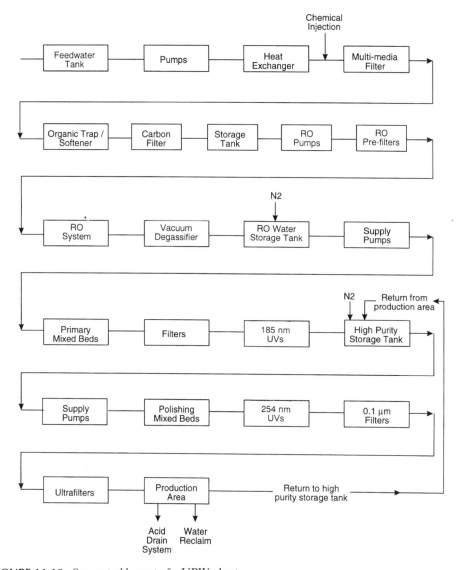

FIGURE 11.10. Suggested layout of a UPW plant.

Instrumentation

The shrinking geometries of successive generations of semiconductor devices has presented serious challenges not only to contaminant removal but also on the development of analysers capable of measuring them at ever-decreasing levels. The days when resistivity was the only measure of water quality have long passed and analysers to detect TOC, particles, silica, dissolved oxygen and individual ions down to ppb levels are now standard.

Continuous monitoring of process conditions is vital throughout an UPW plant. The front end of the plant is more tolerant of changes than the polishing loop but monitoring here is still necessary to detect abnormal variations in the source water. This can be achieved at reasonable cost as the contaminant levels are high and easily detected. However, the cost of providing on-line monitoring increases dramatically when the distribution loop is reached and the cost per point is now approaching £100 000. This cost does not even include the analysis of individual ions which is a major investment in terms of both capital and trained personnel and if these resources are not available it may instead be sub-contracted to a specialist laboratory.

12 The Production and Transmission of High Purity Gases for the Semiconductor Industry

R. Galbraith

INTRODUCTION

The semiconductor industry consumes vast quantities of chemicals, normally in the liquid or gaseous phase. Any chemical which is in contact with the surface of a silicon wafer during the manufacturing process must be of consistent quality and virtually free from all impurities. As the feature size has decreased, and continues to decrease with each generation of semiconductors, the industry has demanded lower levels of impurities on all utilities. Where impurity levels of parts per million (ppm) were once regarded as acceptable, most systems now operate with limits set in the low parts per billion (ppb) or at sub-ppb levels.

The chemicals which are consumed in the greatest quantities are ultra-pure water and bulk gases such as nitrogen, argon, oxygen and hydrogen. Today's specifications for bulk gases extend beyond the impurities which have traditionally been measured, i.e. moisture, oxygen, hydrocarbons and particles. The challenge for today's gas suppliers is not only to produce purer gases but also to develop analysers capable of detecting a wider range of impurities at sub-ppb concentrations.

USES OF BULK GASES

The following subsections describe the design of the bulk gases supply and some of the problems associated with their application.

Nitrogen

The most commonly used gas is nitrogen, which generally exceeds all of the others by at least one order of magnitude. It is generally used as a pseudo-inert gas in furnaces during annealing (usually after ion implantation steps or oxide layer formation) and to control the rate of oxide growth in the furnaces (by acting as a diluent). It is used in chemical vapour deposition systems as an inert backfill when the process is finished and

Cleanroom Design. Edited by W. Whyte © 1999 John Wiley & Sons Ltd

also, in the same equipment, as an inert curtain to exclude air and associated particulate contaminants. In some fabrication areas nitrogen is used during wet processing to provide agitation in tanks containing de-ionized water and other processing liquids. It is used as an air and particulate excluding medium for a variety of equipment, ranging from complex processing equipment (as mentioned above) down to simple boxes containing masks or wafers.

Hydrogen

This gas is used in furnaces in conjunction with oxygen to form steam (the reaction is exothermic and can cause temperature stability problems and quartz damage if the reaction occurs within the furnace tubes and is not tightly controlled). This enhances the rate of oxide growth and hence increases throughput through the furnace. During process steps involving the deposition of aluminium (either in sputterers or evaporators) damage may be caused to the silicon/silicon dioxide interface. The effects of this may be reduced by exposing the wafer to a blend of nitrogen and hydrogen, the latter component of which diffuses to the interface and helps re-establish the bonds.

Oxygen

Oxygen is used principally in furnaces to grow oxide layers on silicon wafers or in conjunction with hydrogen, as mentioned above, for enhanced growth rates. It is also used in chemical vapour deposition systems for film formation. Oxygen is also used in plasma etch systems to strip off photoresist.

Argon

This gas also is used, principally, in furnaces as a more inert (and expensive) version of nitrogen where use of the latter causes processing problems. At high temperatures nitrogen may form nitrides, while at similar temperatures argon remains stable. It is used as a diluent during dopant operations to provide uniform growth rates. If used during the annealing process, it reduces the surface charge at the silicon/oxide interface, giving better control over threshold voltages.

IMPURITIES IN GASES

Obviously, different pieces of equipment have different purity requirements. However, it is not usually economic to supply gases of different qualities and it is therefore normal practice to install a central plant to supply each bulk gas to the quality determined by the most critical process. The main exception to this rule is nitrogen, where it may be financially viable to supply several different grades.

Shown in Table 12.1 is a qualitative list of the impurities which are a problem in bulk gases used in wafer fabrication and an indication as to the reasons for the problem.

The way in which gases are provided within the area of a wafer fabrication facility and how they are conveyed to the piece of equipment in a condition appropriate for use at that piece of equipment should now be considered.

TABLE 12.1. Impurities in high purity gases and their effect on the process.

Carrier gas	Impurity	Effect of impurity
Nitrogen	Oxygen	Upsets control of oxide growth rates in furnaces and disturbs interface levels. In chemical vapour deposition systems, gases such as silane, dichloro-silane and phosphine are used. When exposed to small amounts of oxygen the first two will produce silicon dioxide dust and the latter will also generate particulate contamination.
Nitrogen	Moisture	In furnaces this may, as with oxygen, cause problems at the silicon/oxide interface. The water molecule is smaller than the oxygen one and this fact makes moisture a greater hazard than oxygen with respect to interface problems.
Nitrogen	Hydrocarbons	The presence of these may have two particular effects. First, if they are broken down, carbon may lodge in the oxide layer thus reducing its integrity as an insulator. Secondly, in breaking down it may form water vapour with the same implications as with moisture in nitrogen.
Oxygen	Moisture	As with moisture in nitrogen.
Oxygen	Hydrocarbons	As with hydrocarbons in nitrogen. Also they react with gases such as hydrogen chloride (which are used in many oxide formation processes) to form unwanted products.
Hydrogen	Oxygen	Upsets steam formation rates.
Hydrogen	Moisture	Not much of a problem.
Hydrogen	Hydrocarbons	As per hydrocarbons in oxygen.
Argon	Oxygen	As per oxygen in nitrogen.
Argon	Moisture	As per moisture in nitrogen.
Argon	Hydrocarbons	As per hydrocarbons in nitrogen.

PRODUCTION AND TRANSPORT OF NITROGEN

As already explained, the production of silicon chips requires large volumes of high purity nitrogen and will normally exceed the total of all other gases by a factor of 20–50. This nitrogen may be manufactured off-site at some central facility owned by one of the large gas supply companies or in certain situations on-site. The factors affecting this decision are based on quality, volume, economics and security of supply.

Quality Requirements

The first of the criteria which must be satisfied in the selection process is the capability of the gas supplier to produce the gas quality that matches the process requirements. In the United Kingdom the main gas supply companies are capable of meeting this objective from either on-site or off-site generation. Improved tanker filling and delivery procedures have served to minimize some of the inherent risks of contamination that were formerly associated with imported liquid nitrogen. It should also be considered that on-site generation will normally be backed up with vaporizers using liquid which may come from either liquefaction of site-generated nitrogen or from liquid imports. Therefore to ensure gas quality consistency, the imported liquid must be capable of matching the quality achieved from on-site generation.

Volume

The volume required will strongly influence method of supply but other factors may prevail. If consumption rates are low (say < 500 m3/h) this generally rules out on-site generation. As the volume required increases, the option to produce nitrogen by on-site generation becomes much more attractive in economic terms. There may be other factors which favour the on-site option e.g. availability of local ultra-pure liquid nitrogen imports.

Continuity of Supply

The semiconductor industry has mainly relied on the specialist gas companies to provide their nitrogen requirements through a supply contract. There are several variations on the type of contract depending on site limitations and individual preference. The two components which contribute to the overall supply cost are the rental costs of equipment located on the site and the cost of liquid feed stock. On-site generation requires higher capital investment and on-going maintenance costs on the site, and would therefore result in a higher lease cost. The energy costs of this option are also higher, as the semiconductor company would normally provide power and cooling water for the process. On the other hand, the amount of liquid imports required are minimal and are generally only required during periods of plant maintenance. While the amount of plant which must be installed for off-site generation is considerably less, and this results in a lower lease cost, all the nitrogen used must be supplied from imported liquid. The cost of these imports will depend on the gas company's costs, which include transport, power and their own investment to meet the volumes required.

Nitrogen Generator

A typical layout of a nitrogen generation plant is shown in Figure 12.1. Plants of this type function by liquefying air and then boiling, or flashing off, the different gases by making use of their differences in boiling point. The targets to be aimed at are maximizing quality of product and minimizing quantity of energy consumed (and hence running costs). The principal process steps are as follows:

1. Air is drawn from the surroundings (it being important to locate this intake remote from a source of hydrocarbon rich air, such as a boiler flue) and is fed through compressors to a surge drum. This unit is installed to minimize pressure fluctuations caused by the next component in the process stream.
2. The compressed air then enters the reversing heat exchanger. The function of this unit is to minimize the total energy requirements of the plant by using the cold product and reject gas streams to cool the incoming warm compressed air. The air emerges from the cold end of the unit at about $-160°C$. As the incoming air passes through the heat exchanger, virtually all the moisture and carbon dioxide

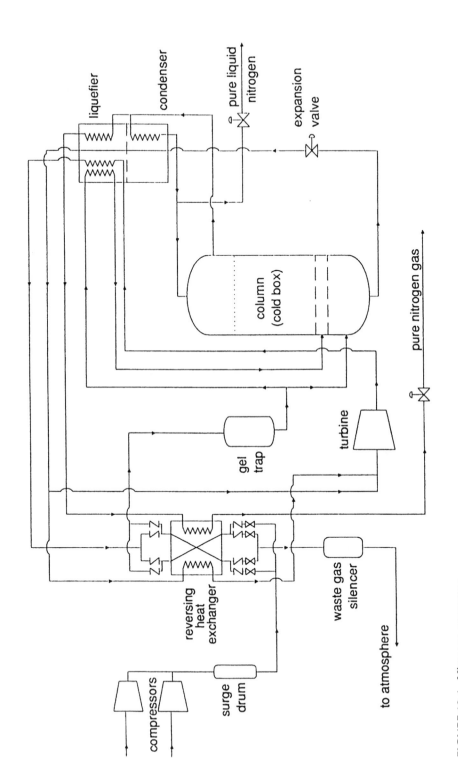

FIGURE 12.1. Nitrogen generator.

in it is frozen out. If this was not so, a fouling factor would develop causing loss of heat transfer (and consequently of the process's thermodynamic efficiency), and also the flow area would reduce with resultant increase in pressure drop with further impact on energy consumption. This build-up problem is overcome by periodically interchanging the flow paths of the incoming air and the outgoing waste gases. The waste stream is able to clean the contaminated passages due to the fact that it is at low pressure and that the volatility of the contaminants increases with decreasing pressure. The relative temperatures of the stream are also important and these are controlled using a turbine and a liquefier/condenser (both described later). This has the effect of cleaning the tubes. The correct frequency of interchange could be accomplished by monitoring the air outlet temperature to determine the precise time when the valves should be changed over, but in practice, for simplicity, the valves are operated on a fixed time base (normally the period would be about 10–15 min).

3. Any remaining carbon dioxide, moisture or heavy hydrocarbons are then removed by passing the cold gas through a molecular sieve.

4. The gas is now a mixture of nitrogen, oxygen and some rarer gases (in very small proportions) such as argon. The gas is now split into two streams, one of which (a) is passed directly to the bottom of the cold box (column). The other stream (b) goes to the liquefier/condenser heat exchanger, where it is liquefied by the cold streams coming from the column. This stream then passes to the column and enters at a location about one-fifth of the way up. The only other stream to enter the column (c) is one consisting of pure liquid nitrogen. This stream enters at the top of the column.

The separation of nitrogen and oxygen is possible due to the difference in their partial pressures. Since nitrogen has the higher partial pressure, the vapour produced from a boiling mixture of N_2 and O_2 has a higher proportion of N_2 than the original mixture. Correspondingly, oxygen will be more concentrated in the liquid phase than in the original mixture. The separation of N_2 and O_2 is achieved by applying this principle in a column consisting of a number of distillation trays. Liquid flowing down the column becomes oxygen enriched, while the ascending vapour becomes increasingly nitrogen enriched. Part of the nitrogen vapour can be drawn off as product, but a certain volume must be condensed and returned to the top of the column as reflux. The N_2 gas obtained from a state-of-the-art plant will contain impurities in the low ppb range and will be suitable for manufacturing processes without any further purification.

The quality now available from on-site generation and also from imported liquid has, over the last decade, shifted preference away from the use of on-site purifiers. There had previously been a desire by some customers to install purifiers that would be capable of providing different grades of nitrogen from a common supply and guaranteeing consistent quality at the process tool. The main disadvantages of on-site purification are the high capital investment (and therefore lease cost) and the potential impact on production of purifier failure. The main advantage is that a lower cost feed gas of poorer quality can be utilized which can result in lower operating costs. This cost reduction will only be realized in practice if the purifier capacity is well utilized as the lease cost of the purifier is fixed, irrespective of the utilization.

A description and diagram of a nitrogen purifier is given in the following section as an example of the technology involved.

Another option which may be specified by equipment manufacturers is the installation of a 'getter' at the point-of-use in the critical process. This is an in-line purifier normally dedicated to one piece of equipment, which is designed to protect the process from any major contamination events. They are non-regenerable and their ability to protect equipment and product depends on their capacity, the impurity concentration and the duration of the event.

5. The oxygen rich liquid is passed via an expansion valve to the liquefier condenser from whence it is split into two streams, the first of which goes to the reversing heat exchanger before rejoining the second stream prior to entering the expansion turbine where the gas is cooled. The purpose in doing this is to provide the appropriate thermodynamic balance to enable the liquefier/condenser to fulfil its principal functions, namely to condense the pure nitrogen gas from the top of the tower to provide reflux and liquid for backup storage, to liquefy part of the incoming air stream and to provide pure nitrogen and waste nitrogen gaseous streams at the correct temperatures to provide the desired thermodynamic balance at the reversing heat exchanger. After passing through the liquefier/condenser the low pressure oxygen rich gas is passed to the reversing heat exchanger where it both cleans the passages of impurities and helps cool the incoming compressed air.

6. The waste gas is then passed through a silencer to atmosphere, the silencer not being required for steady state conditions, but for noise reduction during the period when the high pressure air passage is suddenly depressurized during changeover of the feed and waste streams in the reversing heat exchanger.

The quality of gas obtainable from such a plant would be:

Oxygen:	Guaranteed to 100 ppb
Moisture:	Guaranteed to 100 ppb
Hydrocarbons:	Guaranteed to 100 ppb

If these levels of purity were insufficient for a particular process, a purifier could be used (such as that shown in Figure 12.2).

Nitrogen Purifier

The nitrogen purifier is designed to remove unwanted moisture and oxygen from a process nitrogen gas stream. It is not normal to incorporate any system specifically for hydrocarbon removal as the cryogenic process used to produce the feed nitrogen tends to eliminate virtually all hydrocarbons from it. Hence, although the plant nitrogen is only guaranteed to 20 ppb hydrocarbons, the actual figure is liable to be close to the limit of detection.

Each of the components in the process stream requires to be taken off-line for regeneration and hence the diagram shows two parallel process streams. The process steps are as follows:

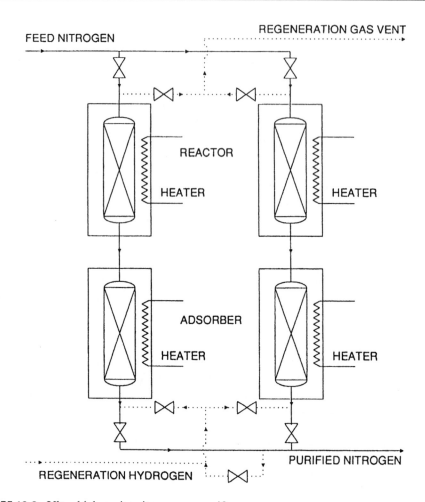

FIGURE 12.2. Ultra-high-purity nitrogen gas purifier.

1. The feed nitrogen enters the reaction chamber which is at ambient temperature. The vessel contains a nickel catalyst which is used to remove oxygen.
2. The gas then passes to the next vessel, the absorber, which contains a molecular sieve. This is used to remove moisture from the gas.
3. When a stream is exhausted the reactor and absorber are removed from line.
4. The reactor and absorber are heated to 250°C and 350°C, respectively.
5. Nitrogen from the on-line purifier is mixed with purified hydrogen in a ratio of about ten to one (by volume). This mixture is passed back through the off-line absorber to remove accumulated moisture and then into the reactor where the hydrogen combines with the oxygen on the catalyst to produce more water vapour. The hydrogen used in this process must be free of impurities such as mercury, lead, iron or any other impurities which might poison the nickel catalyst. This gas is then vented to atmosphere.

The purity levels obtainable from such a system are:

Oxygen:	Guaranteed to 20 ppb
Moisture:	Guaranteed to 20 ppb
Hydrocarbons:	Guaranteed to 20 ppb

In practice, however, levels of purity greater than this may be obtained.

PRODUCTION AND TRANSPORT OF OXYGEN

While plants exist which produce both oxygen and nitrogen from air the relative proportions of each produced do not tend to match the demands of a typical wafer fabrication plant, as in such a plant the demand for nitrogen tends to be orders of magnitude greater than that for oxygen. It is, therefore, normal, even in the case of a large wafer fabrication facility, to import liquid oxygen by tanker to a bulk storage tank and thence, via an evaporator, to the wafer fabrication area.

Two options are available to obtain the required quality at the process tool:

1. Vaporization of ultra-high pure (UHP) liquid oxygen.
2. Vaporization of standard grade oxygen followed by on-site purification.

As with nitrogen, the choice is a balance between the lower operating costs but higher capital costs of on-site purifiers against the low capital-high cost of UHP liquid. The impurity levels for moisture and hydrocarbons which can be typically guaranteed from UHP liquid oxygen are <20 ppb. A description and diagram of an oxygen purifier are given below (Figure 12.3).

Oxygen Purifier

The process steps for the purifier are as follows:

1. The, relatively, crude oxygen is fed via a heat exchanger, into a reactor vessel which is maintained at around 350°C. This vessel contains a platinum catalyst which is used to remove hydrocarbons from the gas by converting them to water and carbon dioxide. This vessel never requires regeneration.
2. The oxygen then passes through a heat exchanger to warm the incoming gases and consequently is cooled.
3. The cooled oxygen then passes via another heat exchanger to reduce its temperature to around ambient and from thence to an absorber.
4. The absorbers (2) are used in an either/or mode. The absorber contains a zeolite adsorbent to remove unwanted moisture and carbon dioxide, and in doing so becomes exhausted (hence the reason for parallel units). To regenerate the unit pure oxygen for the on-line absorber is passed back through the unit (which has been heated to 350°C) and the waste gas and liberated moisture are vented to atmosphere.

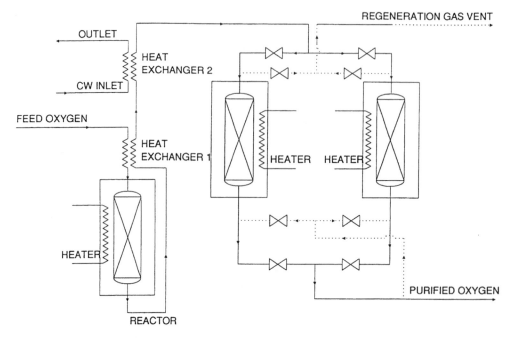

FIGURE 12.3. Ultra-high-purity oxygen purifier.

The quality of gas obtainable from such a purifier is:

Moisture:	Guaranteed to 50 ppb
Hydrocarbons:	Guaranteed to 50 ppb

Similar comments to those discussed with regard to the nitrogen purifier may be assumed in relation to attainable quality levels.

PRODUCTION AND TRANSPORT OF ARGON

It is most unlikely that a semiconductor facility would have an on-site argon manufacturing plant, again due to its relatively small usage. It is, therefore, normally shipped in by tanker in the liquid phase and decanted into on-site storage.

The options for argon are identical to those for oxygen and purity specifications can be met from UHP liquid argon alone. The impurity levels achievable are typically <20 ppb for the main contaminants.

If on-site purification is preferred then a purifier of the type described below could be installed to purify a lower grade feed gas.

Argon Purifier

Such a system is shown in Figure 12.4. The process steps for such a system are as follows:

1. The feed argon is blended with oxygen to give a mixture with less than 100 ppb oxygen. This might seem a retrograde step as oxygen is one of the impurities that we are trying to remove. However, the oxygen is added so that when the mix enters the first reactor (containing a palladium catalyst heated to about 400°C) the oxygen will combine with any hydrocarbons present to form water vapour and carbon dioxide. There is no duplication of the first reactor as it never becomes exhausted.
2. The gas then passes through a heat exchanger to warm the feed gas and be correspondingly cooled and from there to one of two identical streams where it enters the second reactor.
3. This reactor normally runs at ambient temperature and contains a nickel catalyst to remove any oxygen from the gas.
4. The gas then passes to an adsorption unit consisting of a molecular sieve (again at ambient temperature). Moisture, whether from the geed gas or from hydro-carbons in the first reactor, is removed in this unit.
5. When the stream is exhausted it is taken off-line. The adsorber and second reactor are heated to 450°C and 350°C, respectively.
6. Hydrogen, free of catalyst poisoning impurities, is blended with pure argon and back-fed to the adsorber where the argon removes the moisture build-up and from there to the second reactor where the hydrogen combines with the oxygen

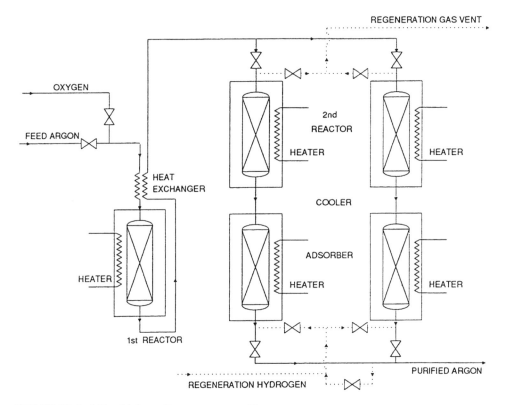

FIGURE 12.4. Ultra-high-purity argon gas purifier.

on the catalytic bed to form water vapour. The waste gases are then vented to atmosphere.

The purity levels obtained from such a system would be:

Oxygen:	Guaranteed to < 500 ppb
Moisture:	Guaranteed to < 500 ppb
Hydrocarbons:	Guaranteed to < 500 ppb

Again, actual levels obtainable are much better than these conservative figures.

PRODUCTION AND TRANSPORT OF HYDROGEN

Hydrogen can be supplied from either a liquid or gaseous source. The type of installation selected depends on criteria such as flowrate, quality required, availability of liquid hydrogen and the associated safety considerations. Gaseous hydrogen must undergo on-site purification to meet process demands and this can be achieved by two methods.

1. The use of palladium cells which permit only the passage of hydrogen because of its small molecular size.
2. A three-stage purifier consisting of a reactor, absorber and a cryogenic stage.

A description and schematic of the latter is given in the following subsection.

Hydrogen Purifier

Hydrogen purifiers are available in two main forms. The simpler of the two utilizes the fact that the size of the carrier gas molecule is less than that of all the impurities in it and consists of a molecular sieve. The second, more complex, type is shown in Figure 12.5. Unlike the first type, it has separate components for removing each impurity. The process is as follows:

1. The crude hydrogen is fed into a reactor containing a palladium catalyst which encourages any oxygen in the stream to combine with the appropriate amount of hydrogen to form water vapour. The reactor operates at ambient temperatures and is not duplicated, as it never becomes exhausted.
2. The gas is then passed to one of two parallel process streams and enters an adsorber. This contains a zeolite based moisture adsorbing medium which removes any moisture from the carrier gas (whether from the original gas or from the oxygen converted to moisture in the reactor). This unit also operates at ambient temperature.
3. The gas from the adsorber passes to a cryogenic purifier which removes any residual moisture and any organics present in the gas. The purifier contains both a zeolite medium and activated carbon. it has been found that this material operates more efficiently at very low temperatures. The material is held in a container surrounded by a cryogenic jacket. This is kept full of liquid nitrogen. The feed and product gases flow through a heat exchanger (both to ensure that the product gas

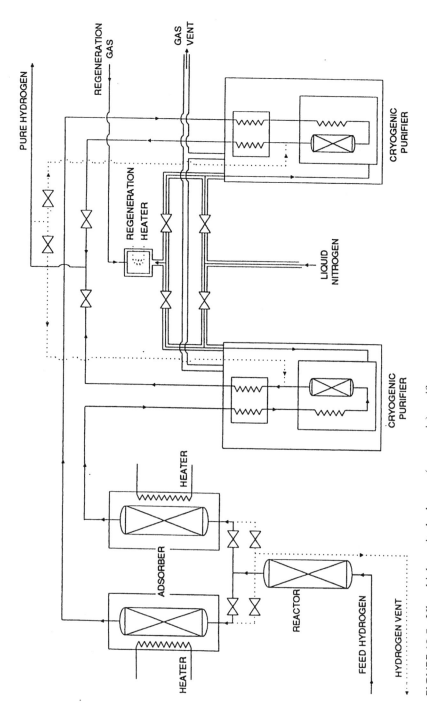

FIGURE 12.5. Ultra-high-purity hydrogen (cryogenic) purifier.

does not leave at a very low temperature and that the feed gas requires the minimum of cooling and hence minimizes liquid nitrogen consumption).

4. When one of the dual streams is exhausted it is taken off-line for regeneration. Nitrogen gas is warmed to 250°C and passed to the reactor and in doing so removes the accumulated moisture and hydrocarbons.
5. Pure hydrogen is fed back through the adsorber (which is heated during the regeneration stage to 350°C) to remove the moisture from it.
6. The waste gases are then vented to atmosphere.

Both types of purifier are capable of producing hydrogen with impurity levels of oxygen, moisture and hydrocarbons of less than 100 ppb.

GAS DISTRIBUTION SYSTEMS

Stainless Steel

The main objective of a distribution system is to deliver the gas from the service pad to the process tool without any degradation in quality. The system components such as pipe and valves must not contribute any particles to the gas stream and must be free from ingress of external contaminants. The system should also be designed in such a way as to minimise dead legs and take into consideration the pressure drops and gas velocity.

The material which is used almost exclusively to satisfy these criteria, as well as providing mechanical strength and corrosion resistance, is 316L stainless steel. A typical analysis is shown in Table 12.2.

TABLE 12.2. Typical analysis of 316L stainless steel.

Element	Level (%)	Element	Level (%)
Carbon	0.03 (maximum)	Silicon	1 (maximum)
Chromium	16–18	Phosphorus	0.045 (maximum)
Nickel	10–14	Sulphur	0.08 (maximum)
Manganese	2 (maximum)	Molybdenum	2–3

A further requirement of the current specification is that any part of any pipe-tube, valve or other fitting which comes into contact with the gas be electropolished and have a surface finish better than 10 Rã.

Electropolishing   Electropolishing is a selective electrochemical metal-removal procedure involving both macro- and micro-polishing (macro-polishing being the erosion of surface flaws, embedded impurities and high points, while micro-polishing is the removal of very small surface irregularities).

When the correct combination of electrolyte, current/voltage, and temperature of the solution is attained, the high points of surface irregularities or areas of high current densities are selectively removed at a greater rate than the remainder of the surface.

By passing a low voltage current through the flowing chemical solution, both macro- and micro-polishing can be made to occur simultaneously producing a smooth highly polished surface. One of the side benefits of this process is that, as it is anodic in nature, oxygen is liberated from the solution and this tends to passivate the surface of the steel producing an even, relatively inert, surface film.

After the electropolishing any chemical residue must be removed by washing with de-ionized water. If ordinary tap water is used some residual staining may occur. It is therefore important that any potential purchaser of 'clean' tube check that his supplier has a small de-ionized facility for the above purpose.

The fabrication method used for stainless steel is orbital welding in which the components are joined together in an inert atmosphere of high purity argon. The parameters for each weld can be tightly controlled and documented to produce consistent welds which are smooth and free from impurities. The utmost care must be taken at all stages of fabrication to exclude contaminants which may be difficult, if not impossible, to remove once the system is complete. This philosophy extends from packaging, inspection, cleaning and drying through to the actual weld procedure and most of these operations take place in clean room conditions.

Valves If the system is made of electropolished stainless steel than it would seem logical to have a valve which is either entirely made of stainless steel or one where all parts of the valve which come into contact with the gas are made of stainless steel. Diaphragm valves are generally preferred to butterfly valves due to the inferior particulate shedding characteristics of the latter. The valve chosen should also have purge ports located either side of it to facilitate both assembly and general purging down of a system, prior to use.

The valves should be supplied by the manufacturer in the following condition:

- Valve internals electropolished to required surface specification.
- Valve body and diaphragm degreased, washed with de-ionized water and dried with argon.
- The cleaned assembly should be packed in a double, heavy gauge, dust free polythene bag, which should be filled with argon and sealed. A label should be stuck onto the bag stating date of packing.

It should be noted that, when ordering the valves, end tubes should be specified which exactly match the dimensions and material specification of the tubes to which it is to be joined. By specifying different pipe, or tube, sizes at inlet and outlet, the valve may also be used as a reducer.

The purchaser should satisfy himself that the cleaning, assembly and packaging processes at the vendor's premises are satisfactory. This is more important with valves than with any other component, as on-site cleaning is impractical, and even inspection of the internals is virtually impossible without generating or introducing contaminants to the inside of the valve.

Fittings All fittings must be made of the same material as the pipe/tube. The fittings must be suitable for orbital welding, that is, the end pieces which are to be welded must have a sufficiently long parallel section to accept the clamps of the orbital welding machine.

They must also have been electropolished to the same standard as the other materials and be supplied cleaned and bagged, the bag being pressurized with argon.

Filtration   Particulate filtration should be incorporated into all gas distribution systems. Ideally the gas should be filtered at entry to the gas distribution system (to minimize particulate accumulation in the pipe line) and also at each user point (as protection against any particulates which may have entered the line or been generated within the system). The filter housing should be made of, and have the same internal surface finish as, the other components used in the system. The filter element should be of pore size 0.1 μm or smaller. Great care should be taken when choosing this filter and, wherever possible, tests should be carried out, using an on-line particulate monitor, to determine the effectiveness of the filter.

Assembly
   (a) *Sub-assemblies.*   These should be fabricated either in a cleanroom or in a room with unidirectional flow hoods to provide cleanroom conditions at the exposed end of the pipe. Class 10 000 or better should be aimed for. All exposed ends should be kept covered when not in this environment. Operatives within the area should wear cleanroom clothing and overshoes and, in addition, protective non-shedding gloves should be worn when handling materials. The general philosophy to be adopted for this part of the operation should be 'it is much easier to keep the internal surfaces clean than it is to clean them up, should they become contaminated'.

   (b) *Assemblies.*   Sub-assemblies should have all open ends covered and sealed in plastic bags before being taken to the location where they are to be welded onto the main assembly. This location should be kept as clean as possible to minimize the likely contamination. In some cases it may be possible to use a portable HEPA filter system when welding sub-assemblies *in situ*. However, space restrictions and proximity of other large diameter service ductwork may preclude this for many installations.

During any welding operation correct purge rates must be established and be allowed to flow for sufficient time to ensure that there are no pockets of air in the vicinity of the area to be joined. It may be considered advantageous to determine this time using a portable oxygen monitor.

Testing   Testing of the completed system should take three forms:

1. Ideally all welds should be X-rayed to check that the standard conforms to the installation specification. If 100% inspection is not carried out (probably for financial reasons) then the most effective technique to be used is probably to carry out a 10% inspection as standard. Where a defective weld is detected, the documentation for each weld should be checked to determine whether there is a common factor in any defective welds found. These may be due to poor clean-

room protocol during sub-assembly or may be confined to *in situ* welds, where conditions are harder to control. They may perhaps be confined to a certain time period, the liquid argon supplied, a welding machine or even to an individual's workmanship. These variables can be minimized to a certain extent by carrying out test welds at regular intervals (minimum start and end of each work period) followed by inspection prior to any further work progressing. Any other suspect welds must be taken out and replaced. The tender documents should be written to reflect this requirement and hence ensure that the costs of such errors are the financial responsibility of the installer.

2. The system should be tested for integrity of components and welds by filling with argon to 1.5 times working pressure and observing temperature-compensated pressure decay over a 24-hour period. If this test is successful, the system is then checked for leaks at each weld following introduction of helium.

3. The third mode of testing relates to ensuring that the installed system is contaminant free. This is done, after purging the system to remove any minor particulate contamination, by passing a high purity inert gas through the system and monitoring moisture, oxygen, hydrocarbon and particulate content at inlet and outlet from the system and, hence, checking for any pickup in the system.

CONCLUDING REMARKS

The foregoing has been presented to give some insight into the complexity of delivering very pure gases (in bulk) to a wafer fabrication area, and while it cannot claim to be comprehensive, all of the major design concepts have been covered.

The bulk gas system which is installed to meet today's requirements must have the flexibility to adapt to the manufacturing demands for purer gases for the following 10 or even 20 years. Significant progress has been made in recent years, but undoubtedly the greater challenge lies ahead for the gas suppliers to meet the industry's exacting specifications, as semiconductors become more complex with smaller feature sizes and therefore more susceptible to damage by contaminants. Indeed, the number of impurities monitored continues to grow and will no doubt extend to some which have not generally been considered harmful in the past.

13 Materials for Services Pipework

T. HODGKIESS

INTRODUCTION

This chapter is devoted to the selection of materials for pipework in high purity water systems. It is presented in two main sections. The first comprises a review of the main candidates for materials of construction for water-treatment plant. The second is concerned with specific pipework components used in the handling of water from the domestic supply through to high purity water. Some attention is also given to pipeline materials for acids and gases.

There are many materials available for the fabrication of pipework and associated equipment. These include *metallics*, ranging from carbon steel through to exotic alloys such as advanced stainless steels, nickel-based alloys and titanium, and *polymeric* materials, encompassing the relatively simple polyethylenes through to expensive fluoropolymers.

This chapter will concentrate upon the following key aspects:

- availability in the required form
- joinability
- cost
- corrosion or degradation resistance

The latter is crucial, (a) because of its obvious relationship to life and reliability of the system and (b) on account of the potential contamination of high purity, product water by corrosion products. A pertinent point in this respect is that high purity water can be unacceptably aggressive to many materials.

METALLIC PIPELINE MATERIALS

Carbon Steel

This material is inexpensive, easily fabricated and assembled (e.g. it has good weldability) and possesses good mechanical properties. A list of mechanical properties of the range of materials discussed herein is presented in Table 13.1. However, carbon

Cleanroom Design. Edited by W. Whyte © 1999 John Wiley & Sons Ltd

TABLE 13.1. Mechanical properties of some metallic and polymeric materials.

Material	0.2% proof stress or yield stress (N/mm^2)	Ultimate tensile stress (N/mm^2)	Elongation (%)	Elastic modulus (MN/mm^2)
0.2%C/5%Mn, C-Steel	195	480	25	210
OFHC copper	60	220	60	110
95/5, Cu/Ni/Fe	120	280	40	
90/10, Cu/Ni/Fe	140	310	40	135
70/30, Cu/Ni	150	380	45	152
70/30, Cu/Ni/Fe/mn	190	460	45	
Type 304 stainless	205	520	35	200
Type 316 stainless	205	515	35	200
20Cr/25Ni/4.5Mo austenitic stainless	220	590		
25Cr/6Ni/1.5Mo duplex stainless	450	800		
HDPE		21–38		
LDPE		10	600	0.1
PP		20–40	70–1100	1.4
PVC (plasticized)		10–24	105–400	
uPVC		35–62	150	2.5–4.1
ABS		30–50	30–45	2.0
PTFE	10	17–39	200–600	0.4–0.6
PVDF	46	39–54	20–150	2.4
CTCFE	31	41–54	200–300	1.7
PFA	15.5	27–32	340	0.7
FEP		20	200–350	0.5
ETFE	24	48	200–500	1.5

Notes: All metallic materials in annealed condition.
Tensile strengths for the polymeric materials refer to 23°C.

steel has very poor corrosion resistance. This disadvantage can be counteracted by the use of coatings such as cement, rubber and other polymers, or with metal coatings (e.g. galvanizing, cladding) but this clearly increases the cost of the final pipe and the reliability of coatings is always a little doubtful.

Copper and Copper Alloys

Copper   This material is widely used in plumbing systems. It has many of the attributes of carbon steel but much better corrosion resistance. However, despite its widespread and generally satisfactory use in domestic/industrial/hospital systems, problems can occasionally arise in certain waters (hot or cold) from pitting corrosion.

Pitting of copper pipes has often been observed to be associated with the presence of films or deposits of various substances. For example, flux residues can promote pitting, and major pitting problems in the United Kingdom about 30 years ago were found to be caused by the presence of a carbon film remaining from the manufacturing process. Various surveys have also reported correlation between pitting tendency and certain aspects of local water chemistry which may be implicated in major problems in Scotland in recent years concerning pitting failures of copper pipework in many large buildings such as hospitals and hotels.

Rapidly flowing water can cause erosion corrosion at bends or at other locations of greater turbulence, such as in the neighbourhood of deposits or at poorly designed couplings. The control of erosion corrosion is primarily a matter of good design and installation practice, involving minimizing sites of potential turbulence, together with control of the overall flow rate. If erosion corrosion problems are experienced when using pure copper pipes, one possible approach is to replace the affected parts by a more corrosion resistant copper-base alloy such as a copper–nickel alloy which would enable the maximum permissible water velocities to be raised from around 2 m/s for pure copper to around 4.5 m/s for 70/30 Cu/Ni alloy.

Copper–Nickel Alloys   These represent materials with significantly superior corrosion resistance to pure copper. Some commercial alloys are listed in Table 13.2; the corrosion resistance increases with nickel content and also, for a given nickel content, with increasing iron concentration.

As indicated in Table 13.1, copper and copper–nickel alloys have significantly lower strengths than many other metallic alloys and this can have repercussions in the design of the pipework systems, especially those transporting high pressure fluids.

The copper–nickel alloys are used in preference to copper for handling more saline waters. The 90/10 alloy is extensively applied in seawater handling systems but not widely used in the water handling systems under consideration here.

Stainless Steels

The term 'stainless steels' encompasses a wide range of alloys containing 11–30% chromium, 0–35% nickel, and 0–6% molybdenum, together with possible small amounts of other elements such as titanium, niobium and nitrogen. The composition of a stainless steel determines the metallurgical structure, and hence influences the properties, of the stainless steel at ambient temperatures. Thus the stainless steel may be ferritic, austenitic, martensitic or duplex (a mixture of austenite and ferrite).

Chromium, the most important alloying element in stainless steels, is a ferrite former. As can be seen in the constitutional diagram shown in Figure 13.1, iron–chromium alloys with more that about 12% Cr remain ferritic at all temperatures down to ambient. On the other hand, nickel tends to stabilize austenite and so, as a very

TABLE 13.2. Some commercial copper–nickel alloys.

Alloy (%)	Ni (%)	Fe (%)	Mn (%)	Cu
95/5, Cu/Ni/Fe	5.5	1.2	0.5	balance
90/10, Cu/Ni/Fe	10.5	1.7	0.8	balance
70/30, Cu/Ni	31.0	0.6	0.8	balance
70/30, Cu/Ni/Fe/Mn	31.0	2.0	2.0	balance

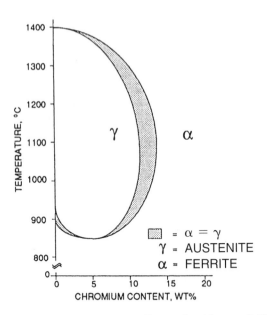

FIGURE 13.1. Iron–chromium constitutional diagram (for steels with up to 0.1%C).

rough guide, increasing nickel content would push the gamma loop of Figure 13.1 further to the right and downwards resulting in the formation of alloys which are fully austenitic at ambient temperatures unless compensated for by simultaneous increases in chromium content. The effects of other elements present in stainless steel, in stabilizing either austenite or ferrite, can be summarized by expressing the composition of the alloy in terms of 'chromium equivalents' (for ferrite promoters) or as 'nickel equivalents' (austenite stabilizers) as follows (compositions in weight %):

Cr equivalents = Cr + 2Si + 1.5Mo + 5V + 5.5Al + 1.75Nb + 1.5Ti + 0.75W
Ni equivalents = Ni + Co + 0.5Mn + 0.3Cu + 30C + 25N

A very approximate guide to the composition ranges of the four types of stainless steels is given below:

Ferritic (12–30% Cr, 0–4% Ni, 0–4% Mo, with low carbon)   The ferritics are not high strength steels (yield strengths in the range 275–415 N/mm^2) but have good ductility and cold formability. They are not amenable to hardening by heat treatment and only slightly so by cold working. Recent developments have led to the availability of low interstitial ferritic stainless steel, i.e. with very low carbon and nitrogen contents (less than 0.04% C+N), with improved ductility, toughness and weldability.

Martensitic (11–18% Cr, 0–6% Ni, 0–2% Mo, 0.1–1% C)   These stainless steels which are austenitic at elevated temperatures but which transform to ferrite upon slow cooling (see Figure 13.1) can, in fact, be transformed into martensitic stainless steels by suitable cooling from the austenitic range. This treatment increases the strength of the alloy (yield strengths in the range 550–1860 N/mm^2) and leads to the use of the martensitic grades in pumps, valves, bolts and bearings. Martensitic stainless steels have poorer ductility and corrosion resistance than the ferritic grades.

Austenitic (17–27% Cr, 8–35% Ni, 0–6% Mo) The amount of nickel required to produce a fully austenitic structure at room temperatures decreases as the carbon content increases. The austenitics cannot be hardened by heat treatment but can be strengthened by cold working. At room temperatures, the yield strengths are in the range 207–1380 N/mm^2 depending upon the composition and amount of cold working. The austenitics also exhibit good ductility and toughness and are generally readily weldable. They usually possess high corrosion resistance, except for a susceptibility to chloride induced stress corrosion cracking (to which the ferritic stainless steels are less vulnerable). In contrast to ferritic grades, nitrogen is beneficial in both stabilizing and strengthening austenite and thus is present at levels up to about 0.2% in some recently developed austenitic stainless steels.

Duplex (18–27% Cr, 4–7% Ni, 2–4% Mo) These have higher yield strengths than the austenitics and were developed to provide materials with higher strength but similar corrosion resistance to the austenitic and ferritic stainless steels. The duplex alloys also often contain up to 0.25% nitrogen and have potential uses as shell and tube heat exchangers and also for pump components and piping in severe environments.

All types of stainless steel are characterized by having considerably superior corrosion resistance in most environments compared with carbon steel. This is due to the presence, on the surface of a stainless steel component, of a very thin adherent, protective layer of chromium rich oxide. This excellent corrosion resistance is a particular feature of all stainless steels and results in their ability to be used, without danger of erosion corrosion, at much higher flow rates than is possible for the copper-based materials discussed in the previous section. On the other hand, stainless steels are rather susceptible to crevice corrosion and to pitting attack in stagnant conditions but the different grades of stainless steel exhibit significant differences in their resistance to these localized forms of corrosion.

As a very general indication, the performance, in respect of resistance to these localized forms of attack, increases with the chromium and molybdenum content of the stainless steel (but unfortunately so does the cost).

Details of a number of commercially available stainless steels are given in Table 13.3 but probably the most commonly used grades of stainless steel for handling all but the most saline of waters are the austenitic alloys with around 18Cr and 10Ni. Two such alloys (using the widely utilized US designations) are the types 304 and 316 austenitic stainless steels with typical compositions indicated in Table 13.3 and in a little more detail below.

Type 304:   18–20% Cr, 8–10.5% Ni, 0.08% maximum C, plus Mn, Si, P, S.
Type 316:   16–18% Cr, 10–14% Ni, 0.08% maximum C, 2–3% Mo, plus Mn, Si, P, S.

The 316 grade, by virtue of its molybdenum content, is superior in corrosion resistance to type 304. Although higher alloyed stainless steels (such as those containing 20Cr/18Ni/6Mo or 29Cr/4Mo) are available, these are significantly more

TABLE 13.3. Some commercially available stainless steels.

Producers	UNS	Other designations or trade names	Nominal analysis			
			Cr	Ni	Mo	Other
Austenitic Stainless Steel						
General	S30400	AISI 304, W.Nr. 1.4301	19	9		
General	S31600	AISI 316, W.Nr. 1.4401	17	12	2.5	
		AFNOR 26CND 17-11				
Several European	S31726	W.Nr. 1.4439	17	13	4.5	N
Several European	NO8904	W.Nr. 1.4539	20	25	4.5	Cu
VDM	NO8925	W.Nr. 1.4529	20	25	6	Cu, N
Avesta-Sheffield	S31254	W.Nr. 1.4547	20	18	6	Cu, N
VEW		VEW963	17	16	6	Cu, N
Ugine		NSCD	18	16	5.5	Cu
General	S31726	W.Nr. 1.4439	17	14	4	N
Allegheny	NO8367	AL-6XN	21	24	6	N
Carpenter		20Mo-6	24	22	7	Cu, N
Avesta-Sheffield	S32654	W.Nr. 1.4652	24	22	7	Cu, N
Ferritic Stainless Steel						
TEW		W.Nr. 1.4575	28	4	2.5	Nb
Trent/Crucible	S44660	Seacure	27.5	1.2	3.5	Ti
Allegheny	S44735	29-4C	29	0.5	4	Ti
Duplex Stainless Steel						
Several Euroopean	S31803	W.Nr. 1.4462	22	5	3	N
Sandvik	S32750	SAF 2507	25	7	4	N
Langley	S32550	Ferralium	25	6	3	Cu, N
Weir Materials	S32760	W.Nr. 1.4501, Zeron 100	25	7	3.5	Cu, N, W

Notes: AISI and UNS are US designations, W.Nr. is a German designation, and AFNOR is a French standard.

expensive than the 304 or 316 grades and are therefore only worth considering for applications involving very saline waters.

It is worth mentioning the modifications to the basic 304 and 316 grades which are available to reduce the susceptibility to 'intergranular corrosion', which can occur in the heat-affected zone of welded joints and in some cast components. Although this is not a problem universal to all types of waters, it is worth considering the specification of stainless steel grades which confer increased resistance to intergranular corrosion. The required modifications to the basic 304 or 316 chemistry involves either using a low carbon grade, e.g. 304L (UNS S30403) or 316L (UNS S31603) which contain a maximum of 0.03% carbon, or the specification of a so-called 'stabilized stainless steel' containing a maximum of either 0.7% titanium or 1.1% niobium.

Surface Finish Issues With Stainless Steels As mentioned earlier, the corrosion resistance of stainless steels depends on the passive Cr-rich oxide film which forms spontaneously in air and most waters. Any event that damages the passive film is obviously likely to lead to corrosion. This can manifest itself as localized pits or more general discolouration—often associated with iron-rich corrosion products (the so-called 'rouging' phenomenon). In pure water systems, the amounts of the metal loss resulting from such corrosion is virtually never a serious problem in terms of the integrity of the component, but it can represent an unacceptable source of contamination of the final pharmaceutical product or electronic device. Fabrication processes can contribute to such corrosion problems both by disturbing and damaging the passive film and also by causing other surface effects and damage:

- Deep scratches along which corrosion can be initiated in service.
- Embedded iron particles arising from the use of steel wire brushes. The iron particles will subsequently corrode themselves and also, by acting as crevice formers, can initiate corrosion on the underlying stainless steel.
- Collection and grinding-in of debris.
- Welding which can leave slag from electrodes, arc strikes, weld spatter and heat tints, all of which can initiate corrosion by damaging the protective film and leaving crevice-forming imperfections.
- Organic contaminants such as grease, oil, crayon or paint markings and adhesive tape can promote pitting or crevice corrosion.

The above possibilities dictate that post-fabrication cleaning procedures should be employed as a matter of good fabrication practice. However, as a safeguard, the purchaser should make them a contractual item. Possible cleaning methods are discussed below.

(a) *Degreasing*. This will remove organic contaminants which themselves can stimulate corrosion or can reduce the efficiency of pickling treatments. Degreasing with non-chlorinated solvents is important since any residual chlorides might induce crevice corrosion or stress corrosion cracking.

(b) *Pickling*. This is very effective in removing embedded iron and other metallic contamination. Pickling is carried out by exposure to a nitric acid/hydrofluoric acid mixture (usually 10% nitric acid/2% hydrofluoric acid). Small objects are best treated by immersion in acid baths at about 50°C. On-site piping or vessels can be pickled by circulation of the acid or by local application of a nitric acid/hydrofluoric acid paste using a paint roller or nylon brush. Either immersion or rinsing with clean water is advisable soon after pickling, otherwise corrosion may be initiated.

(c) *Mechanical cleaning*. If this is used (say by grit blasting, sand blasting, glass bead blasting, grinding) instead of pickling, unsatisfactory results can be produced—such as leaving a highly distorted surface with rough profiles, introduction of contaminants or overheating. To remove such surface defects, a mechanically cleaned surface can be subjected to an electropolishing treatment.

(d) *Electropolishing*. As its name implies, this is a process in which the surface is polished electrochemically by making the component the anode in a DC-electrochemical cell, choosing an appropriate electrolyte (usually phosphoric acid) and voltage/current conditions so as to produce a carefully controlled polishing operation on the component surface. If carried out under optimum conditions, electropolishing produces a satisfactory surface finish from which many types of surface contamination (if previously present) will have been dissolved/removed.

(e) *Passivation*. This is carried out, usually as a final treatment, traditionally with nitric acid (typically 20% strength and warmed). The objective is to thicken the chromium-rich oxide passive film and hence increase the corrosion resistance of the stainless steel. This treatment is ineffective in removing surface contamination but is very useful on machined surfaces. Passivation is sometimes specified after pickling, but is perhaps unnecessary because a pickled surface is already passivated.

Other Metallic Materials

Pipework materials are available with even better corrosion resistance than the stainless steels. Examples are commercially pure titanium and nickel base alloys containing chromium such as Inconel 600 (76Ni/15Cr/8Fe) and Hastelloy X (22Cr/18Fe/9Mo/1.5Co/0.6W/Balance Ni). However, the superior corrosion resistance of these materials is only obtained at considerable extra cost which would not generally be justified for the applications discussed herein.

POLYMERIC PIPELINE MATERIALS

General Comments

The continuous development of new improved polymers over the last few decades has meant that there is an extremely wide range of 'plastics' available for piping systems. All of these consist of giant molecules with a basic framework of carbon and usually hydrogen atoms, but often containing atoms of other elements whose incorporation can result in extensive changes to their properties. Polymers are often subdivided into two classes—thermoplastics and thermosets.

Thermoplastics consist of linear molecules strongly bonded within themselves but with weak bonds between individual molecules. This results in relatively low strength but, importantly from a production and cost viewpoint, yields materials which are generally very amenable to easy fabrication into shapes, such as pipes, by extrusion processes.

Thermosets possess a more three-dimensional molecular structure obtained either by producing materials whose individual molecules are three-dimensional rather than linear or by introducing cross links into a linear polymer structure. Both strategies yield polymers of higher strength which is retained upon moderate heating. Consequently, thermosets are more suitable than thermoplastics for load-bearing applications but are generally more difficult to fabricate.

Most of the polymers used for pipes are thermoplastics but some of the couplings and valves may be made from thermosets.

One characteristic of polymers is their usual high resistance to corrosion (or 'degradation' as it is most usually termed when discussing polymers) in many aqueous solutions of weak acids, bases and salts. However, it should be emphasized that it is dangerous to be too reliant on this general characteristic. Polymers often degrade although by different mechanisms to those involved in the corrosion of metallic materials.

Some polymer-degradation processes can be extremely rapid, leading to essentially instantaneous failure of the component. Such mechanisms can involve (a) direct, rapid *chemical attack* on the polymer, causing destruction of the large molecules by breaking them into short-chain segments, or (b) *Solvation processes* in which the polymer molecules are essentially dissolved. Examples of chemicals which can cause rapid disintegration of vulnerable polymers are strong oxidizing acids. In an investigation undertaken by the author, complete disintegration of a glassfibre reinforced polymer occurred in less than one hour when immersed in fuming nitric acid. Another example is the rapid solvation effect of some organic reagents, e.g. acetone and trichloroethane on PVC. Clearly, the resistance to such rapid chemical and solvation attack by specific agents varies between different types of polymer; this will be discussed later in the chapter.

Other polymer-degradation processes are long-term in nature. One such type is 'hydrolysis' in which $(OH)^-$ ions from the environment substitute on polymer chains. A relevant example is the deterioration of cellulose acetate reverse-osmosis membranes at a pH above 7. Other degradation mechanisms involve 'plasticization' and 'environmental (or stress) cracking'. The former causes a gradual softening of the material and the latter is similar to stress corrosion cracking in metals but can also involve fatigue processes. The mechanism of environmental cracking seems to involve the breaking of intermolecular bonds without softening the material. Reported examples include the effect of NaOH on PVDF (see later). It is well known that the fatigue performance of metallic materials is highly dependent upon the associated environment and this feature can also apply to polymers. However, there appears to be a particular lack of design data for thermoplastic pipelines under fatigue conditions.

It is therefore just as important as it is when utilizing metallic materials to give careful consideration to the chemical resistance of any candidate polymer. In this respect, manufacturers' data on chemical resistance, although often seemingly extremely comprehensive in the sense of the range of environments listed, may sometimes prove to be misleading because such data have often been compiled from tests in highly specific conditions. Thus, as is often stated in such data sheets, this type of information should only be used as a guide, and careful consideration should be given to carrying out a practical assessment of any candidate material in the anticipated service conditions prior to its specification for any purpose. This is particularly true in connection with the behaviour of polymeric components under load.

Another facet of polymer degradation is the possibility of contamination of high purity water by organic constituents which may be rather difficult to detect analytically.

Clearly, the limiting characteristics of polymers, in comparison with most metals, are their lower mechanical strengths and heat resistances. The melting points of polymers can be rather difficult to define and a better indication of the thermal stability of such a material is its ability to support loads at elevated temperatures. The 'heat-deflection temperature' is often used for this purpose; it is determined by noting the temperature at which a certain deflection occurs in a bar loaded centrally. The heat-deflection temperatures of a range of polymers are presented in Table 13.4.

TABLE 13.4. Heat-deflection temperatures of some polymers at 1.8 N/mm (for detailed identification of polymers, see next section).

LDPE	40°C	PVDF	113°C
UHMW PE	43°C	PFA	48°C
PP	49°C	ECTFE	77°C
PVC	60°C	FEP	45°C
PTFE	56°C	ETFE	74°C

Some Polymeric Pipe Materials

Described below are some polymers which are well established for engineering components, including pipework for a wide range of waters, together with other polymers (fluoropolymers) which are too expensive for general application but which have attraction for applications of the type under consideration herein.

Polyethylene (PE)   This is one of the simplest polymers consisting of linear molecules comprising carbon and hydrogen atoms as schematized below:

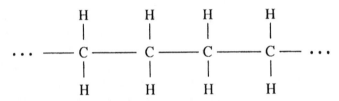

It is widely used in industry for gas-pipework and water-pipework systems, especially where high strength is not required, i.e. where high pressure fluids and/or large diameter pipe systems are not being utilized.

By controlling the polymerization process conditions, PE can be produced with different densities—ranging from 0.91 g/cm^3 (low density, 'LDPE') up to 0.96 g/cm^3 (high density, 'HDPE'). As the density increases, the softening point, tensile strength, abrasion resistance and general resistance to chemicals at room temperature all increase, although the permeability decreases.

Joining of PE pipes and fittings can be carried out by heat fusion, in which pipe ends and/or fitting sockets are heated prior to pushing the two components together.

A special type of PE is the ultra-high-molecular-weight polyethylene, 'UHMW PE'. This has extremely long molecular chains and molecular weights in excess of three million (compared with about 500 000 in the normal PE grades discussed above). This extremely long-chain structure imparts a much greater chemical inertness, including superior resistance to stress cracking, and improved mechanical properties (such as greater impact strength) to the material. However, UHMW PE does not possess the same thermoplastic properties as normal PE grades and components are thus produced by techniques more akin to those adopted for thermosets, such as ram extrusion, compression moulding and final shaping by machining.

All grades of PE are extremely susceptible to degradation by exposure to ultraviolet light such as in sunlight. UV stabilizers (e.g. carbon) can be incorporated into polyethylene; otherwise, when exposure to UV radiation is anticipated, the polymer should be protected by a UV-absorbent coating which should be a solvent-free, water-based paint since some solvents can degrade polymers.

Polypropylene (PP)   This is another widely used material for pipework. Its structure, shown below, contains methyl (CH_3) side groups which stiffen the long chain molecules, imparting a greater rigidity to polypropylene when compared with polyethylene.

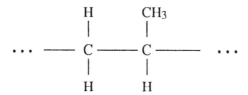

Polypropylene has good abrasion resistance and is suitable for use at higher temperatures than PE. It is usable in pipework up to about 70°C. Like polyethylene, PP can be joined by fusion methods.

Polypropylene has rather similar chemical resistance to PE (with a good general performance at room temperatures) but a lower susceptibility to environmental cracking. PP has good chemical resistance to caustics, some acids and some organic solvents but is vulnerable in oxidizing acids (e.g. fuming nitric acid), low-molecular-weight hydrocarbons and, again like PE, is susceptible to degradation by UV radiation unless suitably stabilized. It is also subject to oxidation attack by ozone.

PVC (Polyvinyl Chloride)

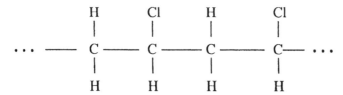

This is another long chain thermoplastic similar to polyethylene but with a proportion of the hydrogen atoms replaced by chlorine atoms.

PVC can be rigid or flexible depending on its composition (i.e. whether it is plasticized or unplasticized).

Rigid PVC (uPVC). This contains no plasticizers. It has a good combination of properties, including high impact and abrasion resistance, is stronger and more rigid than PE (see Table 13.1) and is therefore used for larger bore pipes and higher pressure applications than PE. With regard to chemical properties, uPVC possesses good resistance up to around 60°C in many salts, alkalies and acids with the possible exception of strong acids such as concentrated HNO_3.

Flexible PVC. The properties of this form of PVC are directly related to the proportion of plasticizer, with high concentrations producing soft material with good flexibility and elongation but reduced chemical resistance.

For pipelines, polyvinyl chloride is usually employed in the unplasticized form (uPVC) which is probably the most extensively used polymer for this application. Rigid PVC pipes are made by extruding a mixture of the basic PVC compound plus stabilizer, pigment and toughening additives. They can be easily joined using solvent cements, welding, screwed joints or compression joints. This material becomes brittle below 5°C and is susceptible to surface oxidation and chalking when exposed to UV light.

Chlorinated PVC (CPVC). This is another form of PVC in which the chlorine content is raised from about 55% to about 65% by a post-chlorination treatment yielding a polymer suitable for use at higher temperatures (10–20°C higher than uPVC for a given pressure rating) and higher chemical resistance.

ABS (Acrylonitrile Butadiene Styrene)   This is a polymer based on polystyrene, buta-diene and acrylonitrile. The advantage, in producing such copolymers, is that flexibility is obtained in optimizing the properties of the manufactured material which depend upon the relative proportions of the constituents. In the case of ABS, the styrene provides ease of processing, the acrylonitrile induces chemical resistance and rigidity and the butadiene provides impact strength. Indeed, the latter property is a particular feature of ABS and this good toughness is retained down to sub-zero temperatures (−40°C), making this material particularly suitable for low temperature pipework. ABS also has generally good abrasion resistance, chemical resistance and is non-toxic. However, ABS is attacked by strong acids and is one of the polymers that is susceptible to some deterioration by prolonged exposure to UV radiation.

ABS pipes are easily joined by such methods as cold solvent fusion (utilizing a solvent cement, brush applied, to the pipe and fittings prior to pushing the components together) or gas welding (using an inert gas and a filler rod).

Polybutylene (PB)   This is a semi-crystalline thermoplastic produced by polymeri-zation of butylene (C_4H_8):

$$\cdots \;-\;\underset{\displaystyle H}{\overset{\displaystyle H}{\underset{|}{\overset{|}{C}}}}\;-\;\underset{\displaystyle H}{\overset{\displaystyle C_2H_5}{\underset{|}{\overset{|}{C}}}}\;-\;\cdots$$

Its combination of properties, namely good mechanical properties, similar chemical resistance to PE and PP but with higher UV resistance, and thermal stability, suitability to joining by fusion methods, makes it an attractive material for pipe systems. It is especially attractive for hot-water pipework because of its relatively high upper temperature limit (70°C for continuous use, 85°C for intermittent use). Its long-term temperature/pressure ratings are rather similar to those of CPVC.

Fluoropolymers   This is a group of polymers in which fluorine atoms are substituted for hydrogen atoms. Fluorine is the most strongly electronegative of the halogen elements and thus forms very strong bonds with carbon (much stronger than C bond H

TABLE 13.5. Tradenames and manufacturers of fluoropolymers.

Polymer	Tradename	Manufacturer and country
PTFE	Teflon	Dupont (USA, Holland, Japan)
PTFE	Halon	Allied (USA)
PTFE	Fluon I	ICI (UK, USA)
PTFE	Hostalflon	Hoechst (W. Germany)
PTFE	Polyflon	Daikin (Japan)
PTFE	Algoflon	Montedison (Italy)
PTFE	Soriflon	Ugine Kuhlman (France)
FEP	Neoflon	Daikin (Japan)
FEP	Teflon	Dupont (USA, Holland, Japan)
PFA	Neoflon	Daikin (Japan)
PFA	Teflon	Dupont (USA, Holland, Japan)
PVDF	Kynar	Pennwell (USA)
PVDF	Atochem	Foraflon (USA)
PVDF	Solef	Solvay (Belgium)
CTFE	Kel-F	3M (USA)
CTFE	Diaflon	Daikin (Japan)
ECTFE	Halar	Ausimont (USA)

Note: As indicated above, some companies use the same tradenames to refer to
a number of different materials.

in other polymers) which are responsible for the extreme inertness of the entire class of fluoropolymers. Other features of fluoropolymers that contribute to their superior properties in comparison with other polymers are the absence of chain branching and the relatively high degree of crystallinity. Table 13.5 gives a list of tradenames and manufacturers of commercial fluoropolymers. Some fluoropolymers are fully fluorinated and these possess higher temperature and chemical resistance than partly fluorinated polymers. The latter contain some hydrogen atoms the presence of which increases the intermolecular bonding forces between separate long-chain molecules and this confers superior mechanical properties at room temperature than possessed by fully fluorinated polymers.

PTFE (Polytetrafluoroethylene). This is a fully fluorinated polymer:

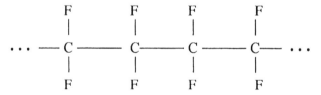

Although of relatively low strength (see table 13.1), PTFE (a) has an extremely low coefficient of friction, (b) has a superior heat resistance over virtually any other commercially available polymer and (c) is chemically inert in an extremely wide range of chemicals, including aggressive acids such as HCl, HNO_3 and H_2SO_4 (some exceptions are alkali metals, chlorine trifluoride and gaseous fluorine at elevated temperature and pressure). It possesses a far greater resistance to a range of inorganic and organic chemicals than any of the aforementioned polymers. It has a wide number of well-known domestic and industrial applications including pipework. However, its very high cost for the latter application certainly restricts its consideration only to transporting 'high value' fluids such as the final purified water in a high purity water system.

Although it is a thermoplastic, PTFE cannot be processed by the conventional melt-processing techniques of normal extrusion or injection moulding. The method generally adopted is to compress resin prior to sintering (at about 330°C) either to the required final shape or a shape ready for machining. Rods, tubes and other continuous shapes can be made by ram extrusion of granular resin.

These manufacturing difficulties mean that despite PTFE's excellent properties it has been necessary to develop fluoropolymers (discussed below) amenable to melt processing.

FEP (Fluorinated ethylene propylene). This is a fully fluorinated thermoplastic with some CF_3 side groups (analogous to CH_3 in PP) which was developed originally to provide a material with similar properties to PTFE but more easily processed. The cheaper processing of FEP is however offset by higher raw material costs. The structure of FEP is shown below.

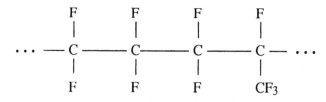

FEP does possess generally similar properties to PTFE with the exception that the latter has a better temperature resistance. FEP also has lower tensile strength.

PVDF (Polyvinylidenefluoride). This is another thermoplastic fluoropolymer which is structurally similar to PTFE and is in competition with it for transmission of high purity water. However, PVDF is not fully fluorinated.

PVDF is another material with an extremely high resistance to many chemicals (including most of the inorganic acids) and also possesses good resistance to UV radiation but is slightly deficient in comparison with PTFE in some aggressive chemicals (because it is not fully fluorinated). Examples of chemicals in which PVDF is not resistant are concentrated hot alkalies, fuming acids, nascent chlorine and chlorinated hydrocarbons.

PVDF has good impact resistance down to low temperatures, good abrasion resistance and is non-toxic. It has relatively high tensile strength and heat deflection temperature.

Joining of PVDF pipework and fittings can be carried out by fusion methods similar to those used for PE systems. Other joining methods are by welding (using filler rods and techniques similar to those used to weld uPVC) and also utilizing compression fittings for small-bore PVDF pipe (up to 12 mm o.d.).

In addition, a modified PVDF-base copolymer has been introduced with lower flexural modulus and is said to be suitable for flexible tubing.

PFA (Polyperfluoroalkoxy). PFA is a copolymer based on PTFE linear molecules which therefore impart many PTFE-like properties including excellent chemical inertness but which promote easier processing, e.g. by extrusion. PFA is reported to possess almost universal chemical resistance at normal ambient temperatures together with a high resistance to environmental cracking. At elevated temperatures it is susceptible to attack in some environments such as molten alkali metals, fluorine and some halogenated compounds. In addition to extrusion, PFA is amenable to injection moulding, welding and thermoforming.

Although usable up to 250°C, PFA possesses lower strength at temperatures above ambient than other fluoropolymers. Indeed, its heat-deflection temperature is the lowest of all the fluoropolymers. Thus, for hot-water piping, it has to be strengthened or used with relatively thick walls. It is also substantially more expensive than PTFE and PVDF.

ECTFE (Ethylene-chlorotrifluoroethylene). This is a thermoplastic containing fluorine and chlorine:

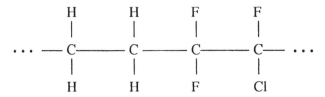

Having a high melting point (240°C) and typically 50% crystallinity, ECTFE exhibits a good combination of properties with good mechanical and thermal stability at temperatures up to 135–150°C and, like all fluoropolymers, resistance to a wide range of chemicals. It has a relatively high impact strength and is said to be less sensitive than PVDF to brittle failure at small notches and other surface imperfections. It can be fabricated by normal thermoplastic processes and has seen application for coating of high-purity water vessels.

ETFE (Ethylene tetrafluoroethylene). With the exception of PVDF, this has the highest tensile strength of all the fluoropolymers but has poorer chemical resistance than PTFE.

SOME ASPECTS OF PIPEWORK DESIGN USING POLYMERS

The very different mechanical and physical properties of polymers (and particularly the thermoplastic materials used for pipes) compared with metals, have consequences for the detailed design of plastic pipework systems. Two such features, discussed below, are the relatively high thermal expansion coefficients and low strength (especially stiffness).

Thermal Expansion Effects

As shown in the data listed in Table 13.6, the thermal expansion coefficients of the polymers considered here are anything from five times to an order of magnitude higher than the metallic materials.

This characteristic can result in significantly greater expansion/contraction of polymer pipe runs compared with metallic systems if the temperature is variable (although the low thermal conductivity of polymers may somewhat dampen the effect of an increase in external or internal fluid temperatures.) Probably the best way of allowing for expansion/contraction effects is by designing pipe runs with regular changes in direction. Failing this, with long runs of pipe, expansion loops can be introduced or, as a last resort, expansion bellows can be incorporated.

Strength/Stiffness Effects

A consequence of the lower strength of polymers is that plastic pipes require more closely spaced supports. As an illustration, the guidelines for support spacing of thin walled PVC, ABS and PVDF pipes filled with water are given in Table 13.7.

Hanger type supports do not provide lateral restraint to polymeric pipes and encourage snaking; a more suitable type of support is shown in Figure 13.2. To avoid

TABLE 13.6. Thermal expansion coefficients (10^{-6}/K) at room temperature.

	Polymers		Metallic materials	
ABS	60–100	Carbon steel		12
uPVC	70	Copper and Cu/Ni alloys		12–17
PTFE and PFA	100–160			
PP	110	Types 304 and 316		
PVDF	120–180	stainless steel		16–17
High density PE	125			
Low density PE	180			

TABLE 13.7. Support distances for plastic pipework.

Pipe o.d. (mm)	Support distance at 20°C (m)			Support distance at 50°C (m)	
	uPVC	ABS	PVDF	UPVC	ABS
12	0.7	0.8		0.3	0.4
25	0.9	1.0	1.0	0.4	0.7
50	1.2	1.2	1.4	0.7	0.8
110	1.8	1.9	2.0	1.2	1.0
225	2.5	2.5		2.0	1.2

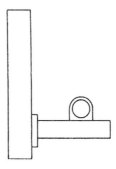

FIGURE 13.2. Example of preferred method of support for polymer pipes.

distortions, heavy valves and metering equipment, etc. should have their own means of support. Metal clips-brackets, etc. should have smooth edges to avoid damage to plastic pipework.

Fibre-Reinforced Polymers

A well established method of promoting increased stiffness and strength in polymers is to incorporate fibres of glass and/or carbon into the polymer matrix. Examples of the improvement of mechanical properties brought about by the incorporation of glass fibres in two polymers are presented in Table 13.8.

There is a wide range of thermoplastic materials and resin systems which can be reinforced in this way, including uPVC and ABS, but probably the polymers most commonly reinforced by glass fibres are polyester and epoxy resins. In certain environments (such as mineral acids) an external coating of pure resin is required to protect the glass fibres from attack. Adequate performance of fibre-reinforced polymer systems up to 150°C is claimed by some manufacturers.

The durability of glass-reinforced polymers (often denoted 'GRP' or 'FRP') is obviously linked to that of the resin component but with the additional aspect of possible degradation of the glass fibres and interfacial attack on the fibre/matrix bond. Attack on the glass component itself is usually only a factor when the material is in

TABLE 13.8. Effects of fibre reinforcement of polymers.

Material	Tensile strength (N/mm^2)	Elastic modulus (MN/mm^2)
ABS unfilled	30–50	2
ABS + 20–40% glass fibre	60–130	4–7
PP unfilled	20–38	1.2–1.6
PP + 25% glass fibre	40–62	3.1–6.2
Polyester resin unfilled	40–90	2.0–4.4
Polyester/glass-fibre woven cloth	206–344	103–310

contact with strong acids. In pure and saline water environments at ambient and moderately raised temperatures (up to about 50–60°C), glass-reinforced polyester, glass-reinforced epoxy and glass-reinforced vinyl ester undergo some reduction in strength but this is usually of a modest extent and occurs in the first few months of exposure followed by a subsequent stabilization of strength. On the question of upper temperature use, suppliers of GRP components sometimes make claims of satisfactory performance at temperatures up to 100–150°C but such predictions should be treated with caution when using conventional resins (e.g. polyester or epoxy).

The main use of GRP in pure-water systems is in the construction of large-volume storage tanks which usually employ the isophthalate type of polyester resin—often with linings of PVDF in order to provide a more inert surface in contact with pure water to minimize contamination by leaching of GRP constituents.

Some Other Design Aspects

Polymeric pipework should be located so as to minimize the chance of mechanical damage or adverse thermal effects, such as might occur, for example, in the close proximity to steam lines. With respect to the latter aspect, a basic characteristic of thermoplastics is their reducing strength with increasing temperature, which means, for instance, that the pressure ratings for pipework must be reduced for higher temperature conditions.

Another mechanical feature of polymer behaviour which must be mentioned is *creep*. This is not a problem with the metallic pipeline materials discussed herein unless they are used at temperatures of several hundred degrees celsius (unlikely in the present applications). However, polymeric materials, because of their lower melting points, are susceptible to creep deformation under load at ambient temperatures and the problems in this respect increase with temperature and may also be dependent on the chemical environment.

Despite their many admirable applications, polymers are often far more susceptible to abuse than are metals. This is particularly true during the assembly of pipework and associated fittings, at which time the polymer components can sustain damage which can lead to poor performance even when the polymer *basically* possesses good physical and chemical resistance for the application in question. Polymers are also more vulnerable to fire damage and may indeed constitute additional health hazards during any outbreak of fires in enclosed areas.

Joining Polymers

Joints in any engineering material are always suspect in terms of strength, durability, leak tightness, etc. but, in pipework for high purity water, additional factors come into play. These are:

- the joint materials, like the pipe itself, must not contain leachable substances which will contaminate the water
- the joint must be smooth and crevice-free to prevent micro-biological settlement, activity and growth.

There are a number of methods available for joining polymer pipes and other components: solvent cementing, threading, mechanical clamping, butt welding, and socket fusion. The details of these processes are easily accessible in the general literature and will not be described herein. Instead, comments will be made concerning their applicability to systems conveying high purity water.

The first three processes listed above are all of doubtful application in this respect on account of the introduction of potential contaminants (e.g. cement material from the first method, lubricants and sealing aids from the second and ferrules from the third). Additionally, mechanical clamping may provide minute crevices. Butt welding and socket fusion are rather similar processes, their difference being that the former applies heat to the faces of the joining ends whereas the latter applies heat to the socket area. Both these processes introduce no other materials and hence no potentially leachable materials, but are said to fail to leave absolutely crevice-free surfaces. For example, butt welding is likely to leave an internal weld bead.

A variant on the butt-welding operation utilizes an internal, air-inflatable bladder positioned in the region of the joint; its function is to maintain alignment of the internal surfaces of the joint during application of heat from the external heating clamp with the aim of producing a beadless and crevice-free weld.

COSTS OF PIPE MATERIALS

Presented in Table 13.9 are some comparative costs per unit length (with the 48 mm carbon steel pipe taken as the reference) for pipes made from most of the materials discussed in detail herein. These are based on UK prices obtained either from published price lists or by enquiry, but it must be emphasized that this exercise was undertaken solely with the aim of providing general indications only of the comparative costs of the different materials. The figures should therefore be treated with caution since material costs can vary significantly from time to time and between different tenders. In this respect, in the main, each price quoted below derives from one source only. Although the data relate to two pipe sizes, namely around 50 and 110 mm o.d., as the list reveals, there are some variations in the precise pipe sizings or ratings pertaining to the different materials.

Notwithstanding the above mentioned qualifications, the figures reveal some clear general cost groupings for the various pipe materials:

1. Uncoated carbon steel represents the least expensive choice but this price advantage no longer prevails if some corrosion protection, by means of galvanizing, is provided for this material. Note also that internal coating by means of painting or with a rubber lining would be considerably more costly than galvanizing for the smaller diameter pipes considered here.
2. One group of polymers, PE, PP, uPVC and ABS, are relatively inexpensive but the fluoropolymers, such as PVDF and PTFE, are around an order of magnitude more costly. Note that although no actual prices are quoted for PFA, pipes made of this material are in fact higher priced than those constructed from PVDF.
3. The other metallic materials listed span the price gap between the two groups of polymers, with 316 stainless steel approaching the price of PVDF. It is worth

TABLE 13.9. Cost of pipe materials.

Material	o.d. (mm)	Wall (mm)	Rating (bar)	Relative (price/m)	o.d. (mm)	Wall (mm)	Rating (bar)	Relative (price/m)
HDPE	50	4.6	10	1.15	110	10.0	10	5.03
PP	50	4.6	10	1.58	110	10.0	10	7.27
uPVC	50	3.7	16	1.42	110	8.2	10	6.79
ABS	50		15	1.88	110		9	5.97
					110		15	7.67
CPVC	50	3.7		3.32	110	8.2		17.09
PB	50	4.6		3.60	110	10.0		14.10
PVDF	50	3.0	16	12.55	110	3.4	10	33.93
					110	5.3	16	49.70
PTFE	50	4.0		23.21	110	5.0		38.67
Carbon steel*	48	3.25		1.00	114	4.5		3.94
As galvanized	48	3.25		1.46	114	4.5		5.53
Copper	54	1.2	27	5.16	108	1.5	17	15.09
90Cu/10Ni/Fe	57	1.5	14	7.45	108	2.5	14	27.15
316L Stainless steel (seamless)†	48	1.65		9.39	114	2.1		29.09

Notes: All prices are for 5 m or 6 m lengths.
*BS 1387 contains 0.20% C max. and 1.2% Mn max.
†Welded 316L stainless steel about 36% of the cost of seamless.

noting that pipes made from the higher alloyed austenitic stainless steel (such as the 20Cr/18Ni/6Mo grade) are significantly more expensive than those made from the 316 alloy.

The prices quoted here are for straight pipes only and do not take into account the cost of fittings and installation. Thus the fully installed costs of a stainless steel system may be nearer those of a PVDF pipeline than is indicated by the above unit length figures. On the same subject, injection moulded PFA components may be much more competitive on price with PTFE for complicated fittings than is the case for simple pipes.

PIPEWORK SYSTEMS

In this section, the choice of materials will be discussed for the different types and parts of piping systems likely to be found in premises where high purity water is being produced.

As-Received Water

This will generally be from the public water system (more rarely from a river or borehole) and will vary chemically from location to location. However, in relation to final product water, it will have the general characteristic of containing a relatively high content of dissolved salts (perhaps 50 to several hundred ppm) organics, suspended solids and dissolved gases. The major criterion in pipework choice will be to utilize a system which gives good reliability at low cost.

The pipework materials for such use have conventionally been carbon steels. Cast iron has also seen some use especially for larger diameter, mains distribution systems. These materials are extremely vulnerable to corrosive attack in many waters especially in rapidly flowing or locally turbulent conditions (such as at pipe bends). In addition to posing maintenance problems, corrosion might complicate downstream water treatment operations on account of the corrosion products introduced into the water. Consequently, for any chance of successful long term use, these materials must be coated. For large diameter, public distribution pipework, internal cement coatings or polyethylene liners are often utilized. For smaller diameter pipes of more relevance in the present context, galvanizing is relatively inexpensive certainly in comparison with coating the pipe internals with paint. However, galvanizing is not always satisfactory for long life.

As shown earlier, some polymers are competitive on price, certainly against coated carbon steel and, in comparison with uncoated steel, are generally less vulnerable to chemical degradation. Consequently, in order to minimize problems associated with corrosion, there is an increasing tendency to utilize polymers in place of ferrous pipework and for this purpose candidates would be PE, PVC or possibly the more expensive but stronger and tougher ABS.

Water Passing Through the Various Stages of Purification

As water progresses through the various parts of the complex treatment train which constitutes a modern high purity water system, it is increasing in value and is becoming more and more susceptible to counterproductive contamination by products of even moderate corrosion or degradation. Thus higher performance pipework materials are more necessary than for the applications discussed in the preceding section. In general, carbon steel or cast iron would not be favoured, although some such steel pipework may be found associated with resin or carbon filters positioned near the front end of the system. Such steel pipework, and also steel vessels comprising the multimedia filters, activated carbon beds and some ion-exchange columns, may be rubber lined. The main candidate materials for the major parts of this pipework are polymers such as uPVC, ABS, PP or GRP and certainly there are many examples of the use of the first two of these materials in high purity water systems, at least up to the final mixed-bed ion exchanger.

However, in locations where higher strength is required, metallic pipes are necessary. One such example is on the high pressure side of reverse-osmosis plants where type 316L stainless steel, on account of its adequate strength and good general corrosion resistance, is commonly employed. Nevertheless, this grade of stainless steel is not always the optimum choice because the critical parameter is often resistance to pitting and crevice corrosion. The latter type of attack is most likely to occur at connections (such as demountable O-ring compression fittings) between the pipe and the pump or the reverse-osmosis cell but may also occur at weld defects.

For a given stainless steel, the prevalence of pitting or crevice corrosion increases with water total dissolved solids (and particularly with chloride concentration) and the resistance to these localized forms of corrosion increases with chromium and molybdenum content of the steel. Work has shown that type 316 (and therefore 316L) is suitable in all but very high chloride waters. In cases where the chloride concentration approaches that of seawater, even the UNS NO. 8904 stainless steel is marginal and a material like UNS S31254 (20Cr/18Ni/6Mo) should probably be used. On the other hand, if low TDS water is being treated, then type 304 (or 304L stainless steel may be sufficient for these applications.

Final Product Water

Durability in relation to external conditions and impermeability to gases from the surrounding environment are relevant factors in the choice of pipes for the conveyance of the high-purity water to, often-distant, points of use. However, of paramount importance is the avoidance of contamination of the water by soluble organic or inorganic species or by local particle shedding. A key aspect in this respect is the polar nature of the water molecule which imparts substantial solvating power especially when it is pure and particularly in relation to ionic substances. Acceptability is restricted to the highest quality metallic or polymeric materials for pipes which must have very smooth non-porous bores to minimize particulate pick-up, shedding and sites for bacterial growth. The maintenance of smooth bores and avoidance of crevices is a particular concern at pipe joints.

Pipe material or jointing aids which contain potentially leachable additives, pigments, modifiers, stabilizers, glues, solvents, sealants, or lubricants must always be suspect. Plasticizers are said to provide nutrients for bacterial growth.

Although PVC and ABS have been used in the past to convey purified water, there are objections to these materials for critical applications where the very highest purity is specified. Thus, PVC is known to release organic (and even metallic) substances present from the production process, ABS may release styrene and contamination may occur from cold solvent joints in pipework involving either of these materials. Polypropylene contains additives such as anti-oxidants as processing aids and these are potential contaminants.

Slight corrosion of a metallic pipe material can put a product water off specification and this factor places a question mark over the medium-cost metallic pipe materials such as Cu/Ni alloys.

To protect the quality of what is a high value product, it is the general practice nowadays for new systems to employ more expensive high performance pipe materials such as stainless steel and the fluoroplastics. In this respect, although 316 stainless steel (especially if pre-passivated to encourage the formation of the protective chromium-rich oxide surface film) may be satisfactory for many purposes and still tends to be favoured for pharmaceutical pure-water systems partly because it is so amenable to steam-sterilization. However, it is not as convenient to assembly and disassemble as polymers and this factor may also close the gap between the final installed cost of a stainless steel system compared with one manufactured in PVDF.

Pharmaceutical Systems

Although PVC has been used in the past and there is some current use of PVDF, the main choice for high-purity water pipework in the pharmaceutical industry is stainless steel. There are occasional instances of concern involving rouging of stainless steel (see earlier) and hence of possible contamination of medical products with Fe, Cr, Ni, etc. but such features are most likely to be accountable to poor fabrication procedures. Factors that account for the continuing acceptance of stainless steel in the pharmaceutical industry are familiarity with its use and concerns about slight reductions in the long-term inertness of PVDF when conveying pure water at high temperatures.

As regards the appropriate grade of stainless steel, the choice is essentially between the 304 and 316 grades, with the latter being most usually favoured when higher resistance to pitting and crevice corrosion are sought. Where welded joints are involved, it is prudent to specify either the low-carbon 'L' grades or the Ti/Nb-containing varieties (see earlier). Also, electopolished bores are often specified for such stainless steel pipes.

Stainless steel is also widely employed for storage of purified water in the pharmaceutical industry especially since hot water (80°C+) storage is often required. Type 316L, sometimes pre-passivated, is the preferred choice for this application.

Microelectronics Industries

Although rare instances of the use of stainless steel or a very expensive polymer, polyetheretherketone (PEEK) are found, the clear trend in recent times, for the final product water distribution pipes and polishing loops, has been the use of fluoropolymers, and especially PVDF. Despite its relatively high cost, PVDF is in fact one of the cheapest of the available fluoropolymers and it is extremely inert, obtainable in the required forms and sizes and can be joined fairly straightforwardly by solventless fusion techniques. Typically PVDF pipework would be utilized from the polishing mixed beds onwards. Other similar materials, such as PFA and ECTFE, may find some application particularly as pipe fittings.

The original shift from conventional polymers such as PVC, ABS and PP to PVDF was founded on experimental studies of comparative leaching rates from candidate materials. The more informative of such studies are those that measure trace substances from samples in a flowing system rather than from static experiments involving exposure of specimens to fixed amounts of water. Some recent investigations of leaching, and other surface behaviour such as susceptibility to biofouling, have compared different fluoropolymers such as PVDF, PFA and ECTFE and different investigations sometimes show apparent similar or slightly superior behaviour of the other fluoropolymers as compared with PVDF. Nevertheless, the overall characteristics of PVDF in terms of mechanical strength, processing and costs, and not least a record of satisfactory performance in practice, have served to maintain the prominence of PVDF for high-purity water distribution pipes.

An alternative strategy, if long pipe runs are involved, is to avoid the excessive cost of materials such as PVDF by utilizing ABS or PVC (materials which have been satisfactory for past and many current specifications) in conjunction with 'point of use' ultrafilters if required to remove any organic pick-up.

Acids

During integrated circuit manufacture, the wafers require treatment with various acids such as sulphuric, hydrofluoric, fuming nitric acid and hydrochloric acid. Another aggressive liquid which has to be handled is hydrogen peroxide. The crucial parameter, in terms of the choice of pipework for such liquids, is the avoidance of corrosion or degradation to ensure adequate reliability of the system and for safety reasons. The lower grade polymers, PE, PP, PVC and ABS, are not generally acceptable for use with many concentrated acids. Although stainless steels of the 316 type or higher grades (such as the 20Cr/18Ni/6Mo steel) can be utilized for handling sulphuric acid and hydrogen peroxide, they are not suitable in hydrochloric or hydrofluoric acid. Consequently, in general, the high grade polymers, for instance PTFE and PVDF, tend to be favoured for this service but with PFA being reportedly particularly appropriate in some instances.

Gases

The handling systems for these are described in detail in Chapter 12 of this book and consequently only a few summarizing remarks will be made here.

From the point of view of deterioration of metallic pipework, there should be no problems since the dry gases involved usually only attack metals at significant rates at elevated temperatures (several hundred degrees Celsius). Consequently the major criterion is the avoidance of contamination of the high purity gases as they pass through the pipes and this factor tends to rule out the use of polymers because of their finite permeability to air.

With regard to copper pipework, which has been utilized together with brass fittings in many circumstances in the past, one potential source of contamination is from flux residues used during soldering and brazing operations and from other films left over from tube manufacture which can be difficult to remove effectively without leaving other contaminants in their place.

Consequently, for modern systems, there has been a tendency to favour stainless steel (such as type 316) with the bore subjected to an elaborate surface treatment involving electropolishing. Although a well established technology for many components, the application of electropolishing to the bore of pipework calls for specialized equipment and expertise and is a costly operation. Similar electropolished stainless steel components are available for valves and other fittings for the high purity gas distribution system.

CONCLUDING REMARKS

It has been shown herein that a wide range of metallic and polymeric materials are available for pipework systems and that the optimum materials for pipework systems handling different fluids will vary.

In general, for water transport, there is a not unexpected correlation between the quality (and hence cost) of the pipe material utilized and the purity of the water. Thus for the as-received water supply and in all but the final stages of purification, the low cost polymers such as PE, PP, ABS and PVC are widely used. But the quality of the final product water is protected by the use of high quality pipework and fittings such as stainless steel or, more usually, fluoropolymers such as PVDF.

The high grade polymers are more versatile than stainless steel for handling a range of concentrated acids but high purity gas lines are more usually metallic, with modern systems utilizing electropolished stainless steel in preference to copper.

However, the proper in-service performance of pipework materials is not simply a matter of the choice of the most appropriate material at the initial design specification stage. Industrial history is littered with examples of piping failures caused by incorrect fabrication and assembly, and the supervision of a piping project from the initiation of the project through to the final installation is necessary.

Index

references to tables are shown <u>underlined</u>; figures are in <u>*underlined italics*</u>